Microelectronics and the Engineering Industry

The Need for Skills

edited by N. Swords-Isherwood and P. Senker

Science Policy Research Unit,
Sussex University

Frances Pinter (Publishers) Ltd., London
Nichols Publishing Company, New York

First published in Great Britain in 1980 by
Frances Pinter (Publishers) Limited
5 Dryden Street, London WC2E 9NW

British Library Cataloguing in Publication Data

Microelectronics and the engineering industry.
1. Occupational training – Great Britain
2. Microelectronics
3. Engineering – Great Britain
I. Senker, Peter II. Swords-Isherwood, N
658.31'24 HD5715.5.G7

ISBN 0-903804-76-X

Published in the USA in 1980 by
Nichols Publishing Co. P.O. Box 96 N.Y. N.Y.

Library of Congress Catalog Card Number 80-82111
ISBN 0-89397-094-8

Typeset by Anne Joshua Associates, Oxford
Printed in Great Britain by A. Wheaton and Co. Ltd., Exeter

Acknowledgements

We are grateful to the Engineering Industry Training Board for sponsoring the research on which this book is based. The book draws heavily on research by several of our colleagues at the Science Policy Research Unit, particularly Roy Rothwell, Ed Sciberras, Mick McLean and Howie Rush. Martin Bell made substantial contributions to the two chapters on welding. Charlotte Huggett, in addition to writing those two chapters made a far more substantial contribution to the book than it was reasonable to ask of any research assistant.

This book would not have been possible without the co-operation of the many people in the industries we studied who gave up so much of their time willingly to contribute data.

Contents

1 Introduction

The structure of the book derives from the development of work at the Science Policy Research Unit for the Engineering Industry Training Board for more than a decade. In 1972, Bell published a pioneering work in which he attempted to assess which new technologies were likely to have a significant effect on the skill and manpower requirements of the engineering industry (Bell, 1972). Bell's work provided a foundation on which we have built the work on engineering processes reported in Chapters 2 (Assembly), 3 (Automated Batch Machining) and 4 (Welding).

Developing this line of research led us to the conviction that, in order to understand industries' skill needs, we must study the characteristics of successful firms compared with relatively unsuccessful firms; identify the critical skills necessary for success; and find out the extent to which these skills are absent in unsuccessful firms (Senker *et al.*, 1976, p.52).

Although the focus of the book is on the United Kingdom, much of the material is of general relevance. Part I contains assessments of the impact of technical change on three important production processes – assembly, automated batch machining and welding. These processes are of crucial importance to the engineering industry worldwide. Chapter 2 considers developments in assembly automation briefly, and points to the differences between the United Kingdom and other countries in the way in which such developments move between the laboratory and the factory. Chapter 3 contains some results of studies of machine shops carried out by ourselves in West Germany as well as in the United Kingdom. These are compared with the results of a somewhat similar study in the United States, carried out by R.T. Lund *et al.* at The Massachusetts Institute of Technology. Chapter 4 examines the importance of welding in the engineering industry and assesses the areas where the application of new types of welding system are likely to be most significant.

Part IIA deals with sectors of the mechanical engineering industry. Chapter 5 examines the position of the British motor industry, but also attempts to assess which of the many possible microelectronic developments are likely to be the most significant in the next few years. With American automobiles now becoming smaller as a result of pressures to reduce fuel consumption, we are moving to the era of 'the world car', in which microelectronics will be significant.

American and Japanese multinational companies are of increasing importance in the British forklift truck industry; trends in technical development in those countries are considered in Chapter 6. European manufacturers are very important in agricultural machinery and textile machinery, discussed in Chapters 7 and 8. American manufacturers are particularly important as suppliers of computers to textile industries. West Germany has long dominated the machine tool industry, discussed in Chapter 9, but Japan seems to be moving fast in the development and installation of very advanced systems. The production and marketing of welding systems is international (Chapter 10). Systems suppliers based in one country often have links with equipment suppliers based in other countries.

Part IIB discusses the skill requirement of electronics and related industries. The convergence of the various sectors of the electronics industries – components, computing, communications, home and office electronics, etc – into 'information technology' is leading also to convergence in skill requirements.

Part III discusses the production of office automation equipment and the likely pattern of office automation over the next few years, focusing on word processing. The 'paperless office' appears to be a long-term, not a short-term, prospect, and there is no evidence at this stage to suggest different patterns of office automation in engineering and other industries.

We emphasize throughout the need to see microelectronics and its skill implications in a broad context. Many chapters build on previous work on the manufacturing processes and industrial sectors considered. In particular, Chapter 3 draws on previous work for the Engineering Industry Training Board and the Automated Small-batch Production (ASP) Committee set up by the Department of Industry at the National Engineering Laboratory. Chapters 6 (Forklift Trucks) and 13 (Minicomputers) also draw on studies sponsored by the Engineering Industry Training Board; as does Chapter 12 (Televisions) which is also based on work by Ed Sciberras for the OECD. Chapters 7 and 8 (Agricultural Equipment and Textile Machinery) are based largely on detailed studies by Roy Rothwell.

The research for Chapter 5 (The Motor Industry), Chapter 9 (The Machine Tool Industry), Chapter 11 (Telephone Equipment), for the two Chapters on Welding (4 and 10) and for Part III on Offices draws on a large number of sources.

The skill implications Chapters (14, 16 and 17) represent an attempt to synthesize the information contained in the rest of the book. But, here again, some specific research, including a number of interviews, was undertaken in order to enable us to draw conclusions. Throughout the book, we draw on a variety of sources, to which reference is made at the end of each chapter.

In view of the inevitable constraints on time and money, many important sectors and production processes are omitted altogether, or covered sketchily. Neither design nor computer-aided design are covered, and nor are the specialized requirements of the semiconductor industry itself. The monitoring of mass production machining, the aircraft and shipbuilding industries are all excluded. We hope to have the opportunity of remedying the more important of these omissions in future research.

Microelectronics components are already widely used in industry and are likely to be used increasingly in the future in ever widening areas of applications. The actual and potential uses of microelectronics cannot all be identified specifically and studied in isolation as can large scale applications of advanced technology such as nuclear power stations or jet aircraft. In some areas – computers for some years, telephone exchanges and office equipment now – the equipments made consist largely of microelectronic components and these components extensively determine the performance characteristics of the products made. In other areas, motor vehicles for example,

TABLE 1.1 Employment in the engineering industry 1973 and 1979

		May 1973 ('000)		*April 1979 ('000)*	
Order VII –	*Mechanical Engineering*	894.2		909.1	
of which :	Agricultural Machinery		24.4		28.2
	Metal-working Machine Tools		62.4		64.1
	Textile Machinery and Accessories		32.9		22.7
	Mechanical Handling Equipment		61.5		59.8
Order VIII –	*Instrument Engineering*	151.3		148.0	
Order IX –	*Electrical Engineering*	746.7		735.8	
of which :	Telegraph and Telephone Apparatus and Equipment		82.2		64.9
	Broadcast Receiving and Sound Reproducing Equipment		57.2		45.8
	Electronic Computers		45.6		47.1
from Order VII	Office Machinery		23.4		22.8
Order XI –	*Vehicles*	782.6		759.6	
of which :	Motor Vehicle Manufacturing		504.6		470.1
	Aerospace Equipment		195.7		198.6
Order XII –	*Metal Goods Not Elsewhere Specified*	514.4		526.2	
		3089.2		3078.7	

Source: Department of Employment Gazette.

a principal activity will always involve mechanical construction, and microelectronics will be somewhat peripheral – but may make significant contributions, for example in terms of safety and fuel economy. Similarly, microelectronic components are being used to enhance the performance of machine tools, mechanical handling and other production equipment. This has further significance: the equipment produced by one industry can transform the production processes of the industries in which the equipment is used. So microelectronic components can affect production processes as profoundly as they can affect products.

There have already been important and useful attempts to assess the long-term effects on society of microelectronics (e.g. Barron and Curnow, 1979). There has also been valuable work on the implications of microelectronics for employment (e.g. Rothwell and Zegveld, 1979).

This book has different, but nevertheless practical and important aims. Advanced industrial economies have to adapt themselves to live with microelectronics. Microelectronic components are cheap and powerful, but they cannot just be 'slotted in' to products and production processes. The most important element necessary for an economy to make effective use of microelectronic components in order to sustain or improve the competitiveness of its industries is the education and training of its workforce at all levels.

The implications of microelectronics appear to be very different for the products and systems produced by the industries considered in Part IIA and Part IIB. The products and systems produced in the industry sectors considered in Part IIA – the motor industry and sectors of the mechanical engineering industry – will need to incorporate microelectronics in control systems to improve product and system performance. But the products produced in such sectors will still be based substantially on other technologies – largely mechanical. Part IIB considers industry sectors in which the role of electronics is central to the functioning of the products and systems produced. Table 1.1 attempts to analyse employment in the British engineering industry in accordance with this distinction.

The skill requirements of engineering industries are of vital importance in the economies of all industrialized countries. Table 1.2 shows that employment in the engineering industries

TABLE 1.2 Wage earners and salaried employees in manufacturing, 1977

	Total manufacturing ('000)	*Manufacture of fabricated metal products, machinery and equipment ('000)*	%
United States	19,647	8,257	42.0
Japan	7,195	3,207	44.6
France	5,419	2,297	42.4
West Germany	8,268	3,899	47.2
United Kingdom	7,352	3,300	44.9

Source: OECD, *Labour Force Statistics, 1966–1977,* Paris, 1979.

TABLE 1.3 Employees in employment in industry (West Germany, France, United Kingdom, 1978

	W. Germany		*France*		*United Kingdom*	
	'000	*%*	*'000*	*%*	*'000*	*%*
1. Mechanical Engineering	990.0	24.8	368.3	15.6	835.1	24.9
2. Instrument Engineering	203.3	5.1	78.8	3.3	150.0	4.5
3. Electrical Engineering	1019.3	25.6	494.0	21.0	704.1	21.0
4. Manufacture of office machinery and data processing machinery	75.2	1.9	44.5	1.9	68.1	2.0
5. Manufacture of motor vehicles	646.9	16.2	487.0	20.7	486.7	14.5
6. Manufacture of other means of transport	151.8	3.8	250.9	10.6	440.4	13.1
7. Manufacture of metal articles (except mechanical, electrical, and instrument engineering and vehicles)	899.0	22.6	633.4	26.9	671.5	20.0
Total	3985.5	100.0	2356.9	100.0	3355.9	100.0

Source: Eurostat, *Employment and Unemployment, 1972–1978,* 1979.

in the United States, Japan, France, West Germany and the United Kingdom represent similar proportions of total manufacturing employment in each country. Tables 1.3 and 1.4 represent attempts to use the available statistics to compare employment in the industries considered in Parts IIA and IIB in terms of their relative importance in West Germany, France, the United Kingdom and the United States.

TABLE 1.4 Employees by industry, United States,* 1978

	Employees '000		%	
1. Machinery, except electrical	2337.3		27.3	
of which:				
Office and computing machines		351.8		4.1
2. Instruments and related products	654.1		7.6	
3. Electric and electronic equipment	1966.5		23.0	
4. Transportation equipment	1956.1		22.8	
5. Fabricated metal products	1652.6		19.3	
Total	8566.6		100.0	

Source: 'Employment and Earnings', US Government Printing Office, March 1979.

*It was impossible to obtain figures for the United States on exactly the same basis as those presented for West Germany, France and the United Kingdom in Table 1.3 above.

REFERENCES

Iann Barron and Ray Curnow, *The Future with Microelectronics*, London, Frances Pinter, 1979.

R.M. Bell, *Changing Technology and Manpower Requirements in the Engineering Industry*, Engineering Industry Training Board Research Report No. 3, Sussex University Press in association with EITB, 1972.

R. Rothwell and W. Zegveld, *Technical Change and Employment*, London, Frances Pinter, 1979.

P. Senker, C. Huggett, M. Bell and E. Sciberras, *Technological Change, Structural Change and Manpower in the UK Toolmaking Industry*, EITB Research Paper No. 2, 1976.

PART I PRODUCTION PROCESSES

This part consists of three chapters on important aspects of some of the main groups of production processes in the engineering industry: assembly, machining and welding. Each chapter includes a brief assessment of the significance of the group of production processes included in the chapter; an account of some of the principal foreseeable technical changes, many of which will incorporate and rely on microelectronic components extensively; and analysis of some of the principal general trends in employment and skill requirements which are likely to result.

As newer electronic components have been developed, and as their costs have fallen continually and dramatically, the range of production process control applications in which they can be applied economically has widened steadily. Our studies of developments in assembly and welding confirm that, despite the rapid increases in the power of microelectronic components, changes in production processes are likely to continue to be gradual in the next few years. For example, there is very intense activity in several countries devoted to development of programmable assembly. It is most likely that this development activity will result eventually in widely used production systems. But long before this happens, the extensive redesign of *products* to incorporate microelectronic components will have more significant effects on manpower and skill requirements.

In 1972, Bell concluded an extenstive survey of the likely implications of production process change in the engineering industry (Bell, 1972). The only area where he considered that production process change would have a dramatic effect on manpower and skill requirements was in relation to numerically controlled machine tools. But subsequent research, reported in Chapter 3, has indicated that, even in relation to numerically controlled (NC) machine tools, effects on skill requirements are likely to continue to be slow.

Chapter 4 reviews developments in welding processes. The significance of this chapter is broader than is immediately apparent as it indicates the implications of technical change for skill requirements in one of the first areas to be affected significantly by robotics.

REFERENCE

R.M. Bell, *Changing Technology and Manpower Requirements in the Engineering Industry,* Engineering Industry Training Board Research Report No. 3, Sussex University Press in association with EITB, 1972.

2. Assembly

P. Senker

THE SIGNIFICANCE OF ASSEMBLY PROCESSES

Assembly is still mainly a manual process. Any attempt to measure its importance in terms of the number of machines in use is liable to underestimate its significance substantially. A better indication is employment. In 1977, about 10 per cent of workers in the engineering industry were semi-skilled assemblers. To this must be added skilled workers engaged on assembly operations – in particular fitters – not separately identified in published statistics. The economic viability of different types of assembly method depends to a considerable extent on the annual volume of units to be assembled. (see Figure 2.1).

Fig. 2.1 The economic viability of assembly methods

	1–300,000	*Annual output range 300,000–3 million*	*1 million and above*
Prevailing assembly method	Manual	Manual	Manual
New alternatives	Negligible	Programmable assembly (being developed: US and Japan)	Fixed Programme assembly (already used to some extent) Automatic insertion of electronic components (Japanese use predominantly)

1. Manual Assembly

Manual assembly is extremely flexible. Many different variations in the product and rate of production are possible. Workers can be retrained rapidly to assemble different products. Each worker can perform several different operations on each product. Manual assembly is more economic than fixed automation assembly for products with smaller annual assembly volumes, or more frequent design changes. The main disadvantages of manual assembly are limited speed and reliability (Lynch 1977).

2. Special Purpose Assembly Machines

At present, the most widely used methods of automatic assembly are of the 'fixed' variety, in which a specialized assembly machine is constructed to make one particular product. A fixed automation assembly machine is really a set of simple machines (with stations) arrayed along a transfer mechanism. Each work station performs a simple operation on the product. Thus, the machine must contain a separate set of equipment for the assembly of each part in the product. The larger the number of parts in the product, the more expensive the assembly machine (Lynch, 1977).

Operations such as machining, welding, applications of adhesives and testing are often included in the sequence. Such special purpose non-programmable assembly machines have been described as 'precision block-stacking machines in which fairly accurate, machined parts are pushed, pulled, or dropped into place in a sequence of steps that may or may not be synchronized' (Nevins, 1976).

Such machines are generally economic only for making at least one million assemblies per year. In some applications, they will show savings over manual assembly with volumes of several million a year. Economic viability depends on a variety of factors – for example, the complexity of the assembly to be made. This affects not only the capital cost of the special purpose assembly machine, but also the time taken to construct the machine and its tooling. If it takes too long to make the assembly machine, then the user is forced to undertake the expense of setting up manual assembly in order to get into production; in addition, the time needed to construct the assembly machine begins to encroach on the life cycle of the product being made, leaving a shorter period in which to amortize capital costs.

Features have been added to such machines, but there have been no important developments involving the application of microelectronics. The use of such machines is increasing, but rather slowly. Potential customers may realize that they can make savings even on the basis of existing output. But, as output has often been stagnant for some time, and as few firms expect substantial increases in output, they are reluctant to install more machines which would have the principal effect of making workers redundant with few prospects of re-employing them.

3. Automatic Insertion of Electronic Components

Automatic insertion is the process of inserting individual components onto a printed circuit board. Prior to the development of automatic insertion, all components were inserted manually and individually into boards. Manual insertion is still widely used; even when automatic insertion is used, often not all components are inserted automatically. For example, components of awkward shapes may need to be inserted manually, or, if a printed circuit board was designed for manual assembly and automatic insertion was adopted subsequently, many of the components may be badly placed for automatic operation and such components may have to be inserted by hand.

The Automatic Insertion Process

The most important machines involved in automatic insertion are sequencers and automatic inserters. Tapes of components in bandolier form are fed into the sequencer. The sequencer rearranges the components in order so that the automatic inserter may subsequently position them as required in the printed circuit boards. Some sequencers stop automatically if components are missing. Some models also carry out some electronic tests on the components themselves.

Subsequently, the components, in correct sequence on the tapes, are transferred to the insertion machines. Automatic insertion machines can generally only handle axial components, although some attempts have been made to develop machines to handle non-axial components. Specialist machines for handling integrated circuits have been developed.

Automatic insertion machines can be reprogrammed to insert different components into different printed circuit boards. So they can be considered as flexible assembly machines in some respects. In order to use them economically, printed circuit boards need to be redesigned to make them compatible with the automatic insertion machines. For example, components need to be arranged in parallel and at right angles only. Further, automatic insertion is only economic when used to process relatively large numbers of standardized boards. So a company's product range may need to be rationalized in order to use automatic insertion economically.

The Importance of Automatic Insertion (AI)

Automatic insertion is particularly important because the redesign of products to use electronic components instead of electromechanical components expands the scope for it in replacing manual assembly of mechanical components and also in place of 'fixed' automation. For example, in 1977, Rotawinder Limited were reported as having set up a plant to manufacture one million reed relays per year on single shift working for the new TXE4 automatic telephone equipment programme (*Automation*, 1977). 'The equipment comprises an automatic tag-inserting machine, an automatic winding and tag-wrapping machine . . . and a special-purpose assembly machine for soldering, metal screen assembly, insulation with tape, testing and finally unloading the finished relays.' Some time in the mid-1980s, when System X has superseded TXE4 this automatic assembly will become obsolete. By this time, lines producing System X modules are likely to be well-established, and these are likely to be printed circuit boards, presumably involving automatic insertion where the volume justifies it.

Areas of Application of Automatic Insertion Machines

Automatic insertion machines have sold to the computer industry, the office equipment industry, the television industry and more recently in telecommunications. The BPO System X programme and associated developments is likely to lead to this being an expanding area for use of automatic insertion (see Chapter 11). In addition, the automotive industry, e.g. initially in fuel control systems, could also provide further areas of application.

Technical Change and Future Developments

In general, the wider use of electronic components in industries which were previously based on electromechanical technology, e.g. in automotive applications, telecommunications and office equipment, could be expected to lead to much greater use of automatic assembly. However, with the progress of component integration – in particular with the wider use of microprocessors – there may be less need for the application of automatic insertion.

Where several printed circuit boards have to be connected up, there is, in principle, an application for automatic assembly and development work is proceeding on this.

4. Programmable Assembly Automation

(a) The Research and Development Activity

Intensive research and development is going on particularly in Japan and the United States. In the United States there is considerable National Science Foundation (NSF) support for research at universities and in non-profit laboratories. Research is also being carried out in such major corporations as IBM, Ford, General Motors, General Electric and Westinghouse, in some cases supported by the NSF.

The key to programmable automatic assembly is continuous feedback of sensory information, so that the machine can adjust continually as need arises, e.g. if it is presented with disoriented components. Much of the research is attempting to deal with such problems. One of the main difficulties is interfacing analogue data recorded from a vision source with digital control machinery (Drury, 1979).

The general consensus seems to be that programmable assembly will be economically viable for items required in quantities of 'a few hundred thousands per year' (See Nevins and Whitney, 1978). Nevins and Draper suggest that, when programmable robots are sufficiently developed, they should be economically attractive for volumes as low as 300,000 per year of one product or group of products, and should remain cheaper than special-purpose machines until the volumes approach 3 million units.

(b) Commercially Available Assembly Robots

A few programmable automated assembly robots are available already. Olivetti uses its own Sigma robot to assemble typewriters and electronic components and claims to be developing applications for other customers. Unimation's PUMA robot, controlled by a microprocessor, is already assembling carburettors automatically for General Motors (Drury, 1979).

EMPLOYMENT TRENDS AND SKILL DISTRIBUTION

Most semi-skilled assembly work takes place in electrical and instrument engineering and in motor vehicle manufacturing. Although they are not separately identified, fitters are also, in effect, skilled assemblers – employed throughout the industry, but heavily concentrated in mechanical engineering. Skilled fitters are not likely to be displaced to any considerable extent by automated assembly, as they generally work on very small batches and the work generally involves diverse skills, some at a very high level.

In total, over half of assemblers are male. Female assemblers are heavily concentrated in the electrical and instrument engineering sectors, which account for nearly three-quarters of female assembler employment, although these sectors account for under 30 per cent of total engineering industry employment. Male assemblers are distributed more evenly through the industry, although nearly half are employed in vehicle manufacture. Table 2.1 shows the employment of assemblers in 1978 by industry sector. Order VII is Mechanical Engineering; Order VIII, Instrument Engineering; Order IX, Electrical Engineering; Order X (MLH370.2), Marine Engineering; Order XI Vehicles; and Order XII, Metal Goods Not Elsewhere Specified.

TABLE 2.1 Employment of assemblers in the British engineering industry, by sector, 1978

	Males ('000)	*%*	*Females ('000)*	*%*	*Total ('000)*	*%*
Order VII	34.9	19.9	12.0	7.9	46.9	14.4
Orders VIII & IX	38.0	21.7	110.0	72.6	148.0	45.3
Order XI	82.7	47.3	13.8	9.1	96.5	29.5
Order XII & Marine Engineering	19.5	11.1	15.8	10.4	35.2	10.8
Total	175.0	100.0	151.6	100.0	326.6	100.0
%	53.6		46.4		100.0	

TABLE 2.2 Assembler employment in the British engineering industry 1973, 1978

	1973 ('000)	*1978 ('000)*	*Net Increase/Decrease*
Male	170	175	+5
Female	184	152	–32
Total	354	327	–27

Most of the reduction in assembly employment between 1973 and 1978 was accounted for by the fall in female employment in the electrical and instrument engineering sectors. The most marked fall occurred in the manufacture of telegraph and telephone equipment where the total number of assemblers, and the numbers of male and female assemblers, were reduced by more than a half. Female workers were affected most as about three-quarters of assemblers in this sector are females. Reduction in employment was caused both by the changeover from the manufacture of electromechanical equipment to the manufacture of reed relay based semi-electronic equipment and by dramatic cutbacks in ordering by the British Post Office. Nearly as many jobs were lost in the manufacture of broadcasting receivers (mainly televisions). This was due to reduction of excess capacity and increased component integration to a greater extent than to automation.

Future Skill Demands

The widespread diffusion of programmable automation in the United Kingdom will face the same problems as has the diffusion of other advanced methods of manufacture in the United Kingdom. In order to get such developments from the research laboratory to the factory, large teams of engineers – including mechanical engineers, production engineers and computer experts – are needed. In Japan, researchers travel with their products into production functions and may well return to research laboratories with years of production experience. In Europe, with each move, the individual expects a substantial salary increase, which inhibits management from moving staff about. By contrast, within a large Japanese company, any movement between functions is part of a normal career within the same company, and the individual who moves does not expect to be treated any differently in terms of status or salary from the one who does not.

CONCLUSIONS

Manual assembly is likely to continue its dominance for many years. In recent years, the principal cause of reduction in the number of female assembly workers employed in the United Kingdom has been changeovers from the manufacture of electromechanical to electronic products, combined with increased electronic component integration. Increasingly, automatic assembly, perhaps mainly in the shape of automatic insertion, may well accelerate assembly employment reduction. Automatic assembly has increasing potential for releasing people from the need to perform repetitive and tedious tasks. This, in itself is to be welcomed. But trade union resistance to technical change may well be greatly increased if unemployment rises in the economy as a whole.

The requirement for intensive effort to get programmable assembly systems into operation will slow up their adoption in all countries. But the experience in Britain with industrial robots in present, established applications, gives further indication that progress in the application of programmable automation will also be slow. Zermeno has estimated (1979) that there were only about 80 industrial robots in use in the United Kingdom in 1978, compared with about 600 in West Germany, 2,500 in the United States and 3,000 in Japan. Zermeno *et al.* point out that adoption of robots is very much affected by managerial skills and incentives. They also point out that international diffusion of robots has been important in the machine tool and automobile industries and the poor performance of these sectors in the United Kingdom (see Chapters 9 and 5) may have slowed diffusion. Widespread failures encountered with the premature adoption of robots in the early 1970s may also have deterred further adoption.

REFERENCES

Automation, 'Automatic assembly and winding on one million reed relays per year', Vol. 12, No. 10, October 1977.

Jolyon Drury, 'Programmable robots – the key to assembly's future', *Metalworking Production*, June 1979.

P.M. Lynch, 'An economic guideline for programmable assembly machines', paper presented to American Society of Mechanical Enginers, 1977.

James L. Nevins, Draper Laboratory, Cambridge, Mass., quoted in 'Automating the assembly line', *Mosaic*, National Science Foundation, Vol. 7 No. 8, September 1976.

James L. Nevins and Daniel E. Whitney, 'Computer-controlled Assembly', *Scientific American*, Vol. 238, No. 2, February 1978.

Ricardo Zermeno, Russell Moseley and Ernest Braun, 'The Industrial use of Robots', in Tom Forester, ed., *The Essential Microelectronics Reader*, Oxford, Blackwell, forthcoming.

3. AUTOMATED BATCH MACHINING

P. Senker

SIGNIFICANCE OF MACHINING OF BATCHES OF COMPONENTS

Metalcutting represents the most important process in the engineering industry. In the Metalworking Production surveys of production equipment in 1971 and 1976, metalcutting machine tools represented 74 per cent and 70 per cent respectively of the total numbers of machine tools and other production equipment listed. These high proportions may overestimate the significance of metalcutting in relation to other processes to some extent. Nevertheless, it is undeniable that metalcutting processes are very important indeed.

The decline in the proportion of metalcutting machines is probably due to a continuing trend in the replacement of metalcutting by metalforming (Bell, 1972, p. 27; Senker and Huggett, 1973, Chaper 2), and to the replacement of larger numbers of less productive manually controlled machine tools by a smaller number of numerically controlled machine tools.

While a high proportion of industry output is produced on mass production machines (from automatic lathes to transfer lines), employment is heavily concentrated in batch production. Total employment in the mechanical engineering industry was 909,100 in April 1979 (*Department of Employment Gazette*, 1979) and most of this involves production in batches rather than mass production. By itself, the mechanical engineering industry accounts for nearly a third of total engineering industry employment. Batch machining processes are also important in many other sectors.

Numerically controlled machining of batches of components (NC) has probably been the most important form of automation in the engineering industry for many years. Numerical control became important even when electronic components much less powerful than microprocessors were available. The advent of powerful but cheap microelectronic components will facilitate the development of more sophisticated and powerful NC systems.

TABLE 3.1 Growth of numerically controlled machine tool installation 1971–1976 and industrial distribution

	Number of machine tools installed	*Mech. engineering*	*Elect. engineering etc.*	*Aerospace*	*Cars, Ships etc.*	*Other*
1971	4,000	1,612	440	760	560	564
1976	9,571	4,452	2,286	1,037	717	1,079
Percentage increase	139	176	420	36	28	91

Source: Metalworking Production's Surveys of Machine Tools and Production Equipment in Britain, 1971 and 1976.

In 1971 the number of numerically controlled machine tools installed in Britain was estimated at around 4,000. Since then the numbers installed have increased quite rapidly to around 10,000 in 1976. The increased use of NC has been particularly important in the electrical engineering, instruments and mechanical engineering industries. Their use in aerospace, vehicles, tractors and shipbuilding has not increased as fast (see Table 3.1). In Britain, as in the United States, the aircraft industry was the first major user of NC. But the use of NC machines has spread widely into all sectors of the industry.

Although the United Kingdom has not yet advanced far beyond the first stage of independent numerically controlled machine tools, the National Engineering Laboratory believe that other industrial nations, particularly the Japanese, are gaining valuable experience from their investments in automated batch production and are planning their first steps towards 'the unmanned factory', as the Japanese describe their goal. Awareness of these trends led the Mechanical Engineering and Machine Tools Requirements Board, Department of Industry, to fund a project with the National Engineering

Laboratory during 1977. The project's aims were

> to study and report on the UK and international activity in small-batch production, to consider its relevance to UK needs in the next decade; to draw up a skeleton plan for the evolution, on a step by step basis, of a completely automated system for small-batch manufacture; and to specify in more detail what should be the research and development aims for the next five years in the UK.

The project was named ASP (Automated Small-batch Production) (ASP, 1978).

TECHNICAL CHANGE AND FUTURE DEVELOPMENTS

Systems such as the ASP fully automated batch production system will require many years before they are completely developed. Before any such system is widely adopted, it will be necessary for companies to take many intermediate steps involving increasing automation of most of the processes of machine shop batch production.

Computer numerical control (CNC) is now in common use in the United Kingdom. It gives additional flexibility. If there are problems, they can be dealt with much more quickly on the shopfloor. But, as yet, there are no true direct numerical control (DNC) or flexible manufacturing systems (FMS) in use in the United Kingdom. These advanced systems are relatively expensive and, compared with other advanced industrial countries where these systems are already coming into use, labour is relatively cheap in the United Kingdom.

However, the widespread use of hard wired NC and CNC systems has resulted in substantial increases in productivity. The advent of microelectronics is leading to dramatic reductions in the number of components requiring to be machined, especially in that part of the industry where products based on microelectronics are superseding electromechanical products. For both of these reasons, the cost of machined components is declining in its significance to the total costs of producing engineering products. This tends to reduce the pressure for technical change in machine shops.

EMPLOYMENT TRENDS AND SKILL DISTRIBUTION

Interviews were carried out in seven machines shops in the United Kingdom in 1970–72 for the EITB (Swords-Isherwood and Senker, 1978a), and in the same machine shops again in 1977–78 (some of these were the same machine shops interviewed as in the West German/United Kingdom comparisons discussed below).

Machine shops making batches of components vary enormously in terms of the type and size of product made, batch quantities, precision standards and a wide variety of other variables. It was clearly impossible within the resources available for the study to interview a sample of machine shops representative in terms of all such variables. However, we consider that the sample of machine shops interviewed was reasonably representative of batch production machine shops in the two countries. The same questionnaires were used in both countries.

In the United Kingdom, it was still the norm for NC machines to be manned by skilled labour. In some areas, skilled manning has been maintained by union pressure. In other areas, semi-skilled workers have been upgraded to skilled when NC machines were installed, again under union pressure. In 1971, it had been expected that operation of NC machines would be undertaken increasingly by semi-skilled operatives, but this appears not to have happened. The balance of skills within machine shops appears to have changed little.

In most shops, employment has been remarkably static. In some, however, drastic reorganization has substantially reduced employment. For example, employment at one establishment in the machine tool industry was 1,800 in 1971. Several NC machines had been installed previously, without causing any redundancies. Subsequently, they were hit by a recession in demand and the labour force was cut back to 1,100. In a textile machinery industry establishment, employment was over 2,100 in 1971 and under 1,200 in 1977. But the principal cause of this reduction of employment was reorganization consequent on loss of share of world markets.

Analysis is extremely difficult because of the large number of variables involved and because of important changes over the time period of the research. In particular, the change in the mix of products produced makes it difficult to arrive at meaningful comparisons of machine shop productivity over time. This is further complicated by reorganizations which result in transfer of work between the various machine shops owned by a particular firm.

PRODUCTION CRAFTSMEN

One of managements' principal motives for installing NC machines has been to maintain and increase production in face of shortages of skilled labour. There is considerable variation in the increase of productivity achievable by installing NC machines, depending mainly on the type of applications in which they are used. However, if one takes it as a 'rule of thumb' that an NC machine produces three times as much output as a conventional machine, then one worker operating an NC machine can produce the output which, without NC would have required three. In general, the installation of NC machines has not affected overall manning patterns substantially. If certain types of conventional machine had been operated traditionally by skilled labour in a particular plant, this pattern of manning has been carried over to the operation of NC machines. The overall statistics confirm that there has been a slow fall in the *proportion* of craftsmen employed in the industry as a whole.

The disturbing feature of Table 3.2 is the decline in *total* engineering employment it demonstrates. This is very largely due to failure to undertake sufficient research and development and to invest in production facilities in order to design and make products which would be more competitive on international markets.

MAINTENANCE, PROGRAMMERS, DESIGNERS, FITTERS

The use of NC machine tools creates a need for new maintenance skills which are in short supply already. The rapid increase in the use of NC machine tools makes it impossible for machine shops to continue to rely on manufacturers for maintenance. Machine-tool manufacturers are trying to alleviate this problem by developing automatic diagnostic equipment based on microelectronics. But the increasing diversity and complexity of control systems is likely to increase the problems

TABLE 3.2 Craftsmen in engineering and mechanical engineering 1970-71 – 1975-76

Employment	*1970–71 ('000)*	*1971–72 ('000)*	*1972–73 ('000)*	*1973–74 ('000)*	*1974–75 ('000)*	*1975–76 ('000)*
Total engineering industry	3,322	3,106	3,127	3,173	3,072	2,891
Craftsmen: number	648	618	605	595	563	546
%	19.5	19.9	19.4	18.7	18.3	18.9
Mechanical engineering*	1,052	951	935	947	927	883
Craftsmen: number†	280	255	247	243	237	226
%†	26.6	26.8	26.4	25.7	25.6	25.6

*The figures relate to the years 1971-1976.
†Estimated. Average of two spot dates.
Source: Engineering Industry Training Board.

of recruiting and training maintenance personnel in the next few years.

Numerical control creates a demand for programmers. It also has indirect effects. For example, the increased precision made possible by the use of NC equipment reduces the requirements for skilled labour in fitting. Redesign of products is also needed to facilitate production on NC machines.

FUTURE SKILL DEMANDS

Below are presented some results from studies in the United Kingdom, West Germany and the United States. First, procedures for deciding on new equipment are compared; then managerial and shopfloor attitudes are considered, followed by consideration of some specific skill shortages in the three countries. We draw on studies in the United Kingdom and West Germany conducted for the Engineering Industry Training Board and the ASP (Automated Small-batch Production) Committee; and on a study conducted in the United States by the Massachusetts Institute of Technology (1978).

The United Kingdom ASP Committee was concerned to establish whether the social and economic barriers to automation were the same in all advanced countries, or whether there were differences between the United Kingdom and other advanced countries. The contraints of available time and budgets caused the

ASP and EITB studies to be confined to the United Kingdom and one other country. West Germany was selected on the grounds that it was a major industrial competitor of the United Kingdom.

At about the same time, a similar study was undertaken independently in the United States by the Massachusetts Institute of Technology (1978), in which twenty-four firms were interviewed. The MIT findings are compared below to our findings in the United Kingdom and West Germany, although it has to be borne in mind that their research aims and methods were somewhat different.

1. Deciding on New Equipment

West Germany

Typically, the initial proposal to invest in new machinery would come from the machine shop manager or the production manager. Graduates would be heavily involved in the initial investigations. New machines were researched and tested and costings and justifications were then prepared and these were submitted to higher management levels which consist mainly of graduate engineers. Proposals had generally to be prepared within a fairly flexible budget. The final decision was mostly made by the owner or managing director. But throughout the process, there was detailed consultation with the workforce. This is a legal requirement.

There were some deviations from this general pattern. At one extreme, the decision-making process used by the manufacturer of heavy industrial equipment was highly complex. Initially, the production manager commissioned machines for trial runs. All those involved or affected observed and monitored the machines and filled in a form which covered many aspects of machine operation and costing. Marks were awarded for every aspect and a total for each machine was arrived at by each individual. An overall total was then calculated. There was nearly always a clear winner, but if there were important discrepancies between assessments, there were discussions and retrials. It was considered that this complex democratic procedure was essential to ensure that machines when installed would be accepted and used well. This procedure ensures that no one individual need feel responsible individually for purchases, which may be very large; and this appeared to overcome the problems of management resistance to new technologies. At the other extreme, decisions about new machinery purchases were made by the two owners of a small family firm employing eighty-five workers which made some proprietary machines and also did sub-contract work.

The United Kingdom

Decision-making procedures were broadly similar in the United Kingdom. The most significant differences to emerge were that, partly as a result of the West German legislation, participation of the workforce in West Germany tended to be more formal and official and in the United Kingdom more informal and friendly; and engineering graduates were involved to a greater extent in decision-making in West Germany than in the United Kingdom. In the United Kingdom, some companies had relatively elaborate decision-making procedures. But none were as elaborate nor involved as many people in such depth as the extreme West German case. The British subcontractor's decision-making processes were broadly similar to those in the comparable West German firm, being based on discussions between the owner, the works manager and the foreman.

The United States

At many of the firms, it was found that there was a lack of financial data. Equipment justification did not appear to be performed using well-developed cost projections backed up by historical data. It was rare for firms to compare their actual experience with that they had predicted before purchasing the equipment.

2. Managerial Attitudes to Numerical Control

West Germany

Initially, managements had resisted NC. Unreliability caused problems and generated resistance amongst supervision and managements in some firms, as did ignorance about NC and about its possible consequences in terms of reorganization. In one firm, the management had been worried initially about responsibility involved in the large expenditure necessary to invest in NC. These worries had been overcome by the time of the interviews (1977).

The United Kingdom

In general, managerial attitudes to NC were not quite as favourable as in West Germany. Relatively greater emphasis was given to cost rather than engineering considerations in the United Kingdom. Middle management in a machine-tool firm considered that there was resistance to NC amongst top management. The other machine-tool firm considered that it lacked sufficient managerial and engineering expertise to get the best out of NC equipment.

The United States

The MIT study concluded that NC was still something of a mystery to many middle managers and supervisors. As in the United Kingdom, many American supervisors had been machinists prior to promotion. As machinists, they had not used NC machines and so were unfamiliar with them. They were uncomfortable in their heavy dependence on technical support specialists. But American managers were generally aware of the benefits of NC in terms of more consistent quality, lower scrap losses, faster production times and lower set up and adjustment time. Some felt that it was easier to plan machine workloads for NC equipment, as machine capacity was more predictable.

3. Shopfloor Attitudes to Numerical Control

The interviews were conducted with managements in both European countries. Thus, it would be reasonable to expect that the resistance of the workforce would be exaggerated. It is remarkable that managements in neither country considered that workers or trade unions had impeded the progress of NC significantly. Indeed, this was another area where competent management could minimize problems. In both countries, managements thought that NC machines reduced skilled men's job satisfaction substantially. Exceptionally, in one company where semi-skilled workers were employed operating NC machines, they were welcomed as making jobs more interesting.

The application in which the NC machine is used may be

critical. For example, the operation of an NC machine in a shop where complex one-offs and very small batches are made (e.g. in prototype production or precision toolmaking) may provide considerable job satisfaction. In some cases, skilled workers have been very keen to work on new machines. In other cases, NC may lead to jobs offering considerably reduced satisfaction where larger batches of simpler components are being made. The advent of microprocessor control systems gives management the possibility to build in more opportunities for shopfloor intervention to improve performance. Managements may be reluctant to take advantage of this because it would lessen *their* opportunities for controlling output.

West Germany

One company, the manufacturer of heavy industrial equipment, had encountered shopfloor resistance to NC. The works council was not as against NC as the individual shopfloor workers. Whilst this was the only West German company interviewed where actual shopfloor resistance to NC had been encountered, NC was unpopular amongst skilled shopfloor workers in five of the six machine shops. Skilled men felt that they had less chance of using their skills than on conventional machines.

These firms differed from several of the British firms as they had not recently experienced large-scale redundancies. There were fears of workforce resistance to NC in the future.

The United Kingdom

In Britain, there had been no significant resistance to the installation of NC, although in several plants there had been large redundancies. This may be because, in general, there was no direct and obvious connection between the installation of the NC machines and subsequent redundancies. There had been some initial problems with NC in some firms – described by one firm as 'scepticism about NC' rather than resistance to it. In another firm, there had been difficulty in devising equitable payment schemes as there was less opportunity with NC for the worker to earn more by working harder.

One firm pointed out that, although the effects of the

installation of NC on industrial relations had been negligible, if acquisition of NC had not been well planned and preceded by discussions with the shopfloor the effects could have been disastrous.

The United States

It was found in the MIT study that the firms interviewed had experienced no significant resistance to the introduction of NC from their workers. Any resistance experienced was not considered as deterring future NC acquisition. But the researchers remarked that NC machines appeared in a period of growth for most firms in their study. In future, if the climate is less favourable, resistance may be encountered.

4. Skill Shortages

West Germany

Two of the West German companies interviewed were experiencing skill shortages at the time of the interviews (1977). Both were in the most heavily industrialized area in Germany. One was overcoming this by training *gastarbeiter*. Training these men, who came from rural areas, was a big job for production engineers and *obermeisters*.

The United Kingdom

Skill shortage problems were somewhat more widespread in the United Kingdom. Four of the firms interviewed had problems in obtaining sufficient craftsmen. This could possibly be partly because the demarcation between craftsmen and semi-skilled workers is not so closely controlled in West Germany as in Britain. For example, a German firm which suffered craftsman shortages was able to solve the problem by upgrading semi-skilled workers relatively quickly; this would not normally be possible in British firms.

In the United Kingdom, trade unions have tried mainly to secure higher wages. Possibly the reduced job satisfaction experienced by *craftsmen* called on to operate advanced batch-machining systems may arise because they are likely to be more concerned about job satisfaction than semi-skilled workers (e.g. see Goldthorpe *et al.*, 1968, pp. 24, 25, 83, 84). This may be a factor which will cause difficulties in the supply of craftsmen in the future; even though the increased productivity of automated batch production systems could combine with continuing decline in the British mechanical engineering industry's competitiveness to reduce the overall demand for craftsmen.

The United States

The MIT study did not identify shortage of skilled labour as a problem area. But although recently installed NC equipment was found to be reasonably reliable, when it did go wrong, it took a long time to find faults; once found, the faults could be fixed relatively quickly. Many people were normally required for diagnosis, but there was no reported shortage of such people.

CONCLUSIONS

It was in the area of attitudes to graduate engineers, their recruitment and deployment, that the greatest difference was found between West Germany and the United Kingdom. In the United Kingdom, graduate engineers were to some extent regarded as a luxury. At best, it was felt that recruitment and training of graduate engineers was a long-term investment with a speculative pay-off. Graduate engineers were thought generally to be no better than non-graduates, but more expensive to employ and train. There was some preference for sponsoring apprentices on sandwich courses. Familiarity with the company, its products and customers was rated more highly in Britain than technical expertise.

British management were looking for their engineers and managers among ex-apprentices, with detailed shopfloor knowledge, who acquire technical knowledge in-depth through part-time training and sandwich courses. This route has supplied many good engineers in the past and doubtless will continue to do so in the future. However, it is more doubtful whether this route can supply enough engineers to meet future needs. An extreme example of this was the firm which needed first class practitioners in all the major areas into which they were introducing NC – turning, milling, boring. They needed people who were capable of taking a broad look at the total manufacturing system, and felt that these people needed to emerge from the ranks of the time-served craftsmen. This has to be contrasted with one of the German firms – of smaller overall size – which had no such problems and employed thirty-five engineering graduates.

Nearly all the West German companies recruit graduate engineers and technologists as a matter of routine. A typical view was 'We make very specialized and technically sophisticated machines and therefore we need highly qualified people!' They believed it essential to have such people, for example, in order to bring in new ideas. They expected graduates to be useful shortly after they were recruited. Graduates were not found more difficult to cope with than any other category. If they were good, they progressed into management. Only the small sub-contractor had no policy of recruiting graduates and employed none. None of the firms complained of a shortage of engineering skills.

At present, resistance to automated batch machining arises partly from managements' reluctance to accept increased responsibility. This seems to be a bigger problem in the United Kingdom and the United States than in West Germany. In West Germany, technical and economic justification of new machinery is carried out in great detail in all but the smallest companies and engineering graduates play an important part in these evaluations at all levels, up to and including top management. In the United Kingdom, evaluations are not, generally, so thorough, and involve a higher proportion of part-time trained engineers.

One of the main reasons for the greater unwillingness of British companies to employ engineering graduates is because these graduates often need extensive training before they can play a useful role in industry. In Germany, this is not so. German managements reckon that the graduates emerging from their technical universities can be useful very soon after they are recruited.

The continuing shortage of appropriately educated, trained and experienced manpower, in particular production engineers, makes it likely that technical change in British machine shops will continue to be relatively slow.

In West Germany, the United States and the United Kingdom, managements are concerned that union and shopfloor fears of redundancy could lead to increased resistance to automated batch machining in the future. Wider use of automated batch production systems could lead to reduced job satisfaction amongst production craftsmen. In the United Kingdom the increased productivity of automated batch production, and possibly continuing decline in international competitiveness, are likely to lead to reduced demand for production craftsmen. But low differentials and declining job satisfaction could combine to reduce the supply of production craftsmen, leading to continuing periodic shortages.

Microelectronics makes possible the development of more sophisticated batch production systems which are likely to require less labour for their operation. Reduction in the quantities of components needing to be machined, combined with the slow diffusion of more advanced machining methods, are also likely to alleviate craftsmen shortages, although local shortages will doubtless continue. It is possible, however, that manufacturers' attempts to make sophisticated systems reliable and easy to maintain may not be entirely successful, so the requirement for maintenance workers could increase.

REFERENCES

Automated Small-batch Production, National Engineering Laboratory, Department of Industry, 1978.

R.M. Bell, *Changing Technology and Manpower Requirements in the Engineering Industry*, Engineering Industry Training Board Research Report No. 3, Sussex University Press in association with EITB, 1972.

Department of Employment Gazette, July 1979.

John H. Goldthorpe, David Lockwood, Frank Bechhofer and Jennifer Platt, *The Affluent Worker: Industrial Attitudes and Behaviour*, Cambridge University Press, 1978.

R.T. Lund, 'Numerically Controlled Machine Tools and Group Technology: A Study of US Experience', Center for Policy Alternatives, Massachusetts Institute of Technology, 1978.

P. Senker and C. Huggett, *Technology and manpower in the UK engineering industry, An Interim Report*, EITB Occasional Paper No. 3, 1973.

N. Swords-Isherwood and P. Senker, 'Technological and Organisational Change in Machine Shops', Science Policy Research Unit, mimeo, May 1978a.

N. Swords-Isherwood and P. Senker, 'Social Implications of ASP', Chapter 2 in *Automated Small-batch Production*, National Engineering Laboratory, Department of Industry, 1978b.

4. Welding

C. Huggett

THE SIGNIFICANCE OF WELDING PROCESSES

Welding accounted for about 6 per cent of metalworking production activity in the engineering industry in 1976 (see Table 4.1) and seems to have been growing in relative importance since 1971 when, on the same basis, it constituted only about 4 per cent of metalworking production activity.

TABLE 4.1 An approximate indication of the relative importance of welding in the engineering industry

	1971	*1976*	*% change*
Number of welding machines	40,095	60,965	52%
Number of welding machines as % of all metalworking machine tools and equipment	4.1%	5.8%	41%

Source: Metalworking Production Survey of Machine Tools and Production Equipment, 1971 and 1976.

Welding activity is heavily concentrated on resistance welding – (see Table 4.2). Arc welding equipment and 'other' fusion welding equipment accounted respectively for about 12 per cent and 24 per cent of all welding equipment. Other categories, including most of the more 'novel' forms of welding such as ultrasonic, laser, electron beam etc., accounted in total for only about 14 per cent of welding equipment (Huggett and Bell, 1979).

TABLE 4.2 An approximate indication of the relative importance of different types of more mechanized welding process

	1971 *Number*	*% of all welding machines*	*1976* *Number*	*% of all welding machines*
Resistance welding All categories, eg. spot, seam, flash butt, projection, percussion	23,479	58.6	30,732	50.4
Fusion welding Arc welding (not hand)	4,003	10.0	7,068	11.6
Other (excl. electron beam and laser)	8,344	20.8	14,517	23.8
Other welding inc. pressure, ultrasonic, friction, laser, electron beam and gas welding machines (not hand)	4,269	10.6	8,648	14.2
Total	40,095	100.0	60,965	100.0

Source: Metalworking Production Survey of Machine Tools and Production Equipment, 1971 and 1976

The importance of welding relative to other forms of metalworking production in different sectors of the industry is shown in Table 4.3

TECHNICAL CHANGE AND FUTURE DEVELOPMENTS

For some years electronic devices have been incorporated in welding systems. These devices have been used to effect, or to assist with, the various control functions involved in the welding process. Developments in microelectronics technology are giving rise to yet cheaper and more reliable basic devices. These developments present opportunities for more widespread use of existing types of control system, and for the development of more elaborate and complex devices which incorporate within electronic systems various types of control function hitherto effected by other means – or not at all.

Electronic systems are used to carry out the control func-

TABLE 4.3 The relative importance of welding within different sectors of the engineering industry

MLH		*Welding equipment as a proportion of all metalworking machinery* %
394	Wire and wire manufacture	20.3
341	Industrial plant and steelwork	10.7
380/81	Tractors and motor vehicles	8.5
331	Agricultural machinery	7.8
336	Construction equipment	7.6
363/64/65/66/67	Radio, telephone & electronic equipment	7.4
384/85	Railway locos. & rolling stock	7.3
333, 338/39	Pumps + office and other machinery	6.7
399	Metal industries nes	6.7
337	Mechanical handling equipment	5.7
368	Domestic electric appliances	5.7
361	Electrical machinery	5.4
369/362	Other electrical goods, wires & cables	4.0
351/53/54	Instruments + photo & copy equipment	3.6
370	Shipbuilding and marine engineering	3.4
395	Cans and metal boxes	3.4
335	Textile machinery	3.2
383	Aerospace equipment	2.3
349	Other mechanical engineering nes	2.2
391	Hand tools and implements	2.1
390	Engineers' small tools and gauges	1.2
332	Metalworking machine tools	0.9
334	Industrial engines	0.7
393	Bolts, nuts, screws, etc.	0.4
	Average	5.3

tions associated with welding production:

(a) control of the basic welding process parameters;

(b) control of the movement and location of the welding device in relation to the workpieces;

(c) control of (i) the movement of workpieces to and from the work station, and of (ii) the orientation and location of workpieces within it – in relation to each other and to the welding device or to fixed datum points.

This equipment forms only one part of the technical system needed to carry out the welding process as a whole. There is adequate evidence to indicate that the dramatic economic implications of improved electronic systems *per se* become substantially 'watered down' when they are set in the context in which they are used.

Much of the relevant information needed to control the welding process must be derived from the welding area itself. Irregularities in workpieces may require flexible control of the movement of the welding device, or changing operating conditions may require frequent adjustment to the process parameters. The eyes, ears and expertise of human welders can cope quite well with such problems. Machine systems need to be equipped with comparably adequate 'sensors' and stores of decision rules of various types if they are to do so. In many cases the necessary types of sensor have not been developed, and where they have been (as with some 'tracking' devices for arc welding) these add to the cost (and often the 'non-electronics' cost) of the whole system.

Developments in electronics technology are being incorporated in a very wide range of different kinds of change in welding production systems.

1. Systems which Incorporate Electronic Controls together with the Welding Device Itself and a Robot-type Manipulator of the Device

Initial application of such systems was heavily concentrated in resistance welding where control requirements were usually less demanding than in arc welding. In addition, application was concentrated in areas of high volume production such as the automobile industry.

Despite the programmable versatility of this category of system (i.e. robot control without mechanized handling), we would expect relatively limited application to lower-volume areas of production in the near future, unless basic economic conditions quite substantially shift the relative costs of capital and labour.

The use of robot welding systems in arc welding applications has so far been relatively limited. To a significant extent use has been confined to applications involving 'unpleasant' and 'awkward' working conditions – situations where more narrow economic calculations are complemented by other considerations. One additional factor has contributed to limiting the extent of application of robot systems in arc welding. Very many types of application require more precise and adaptive locational control and to some extent more complex control of process parameters than in the case of resistance welding, and mechanisms have not yet been developed to deal wholly and effectively with this problem.

Without such 'adaptive control' features, the application of robot-controlled arc welding is likely to be heavily concentrated in areas of high-volume production, where high levels of workpiece quality standardization can be more easily achieved, and where mechanized handling systems can more easily ensure more consistent location and alignment of workpieces.

2. Systems which Integrate Mechanized Workpiece Handling with Electronics-based Robot Operation of Welding

Systems of this type have been used for resistance welding for some years – mainly in the automobile industry and they are relatively common in automobile plants in continental Europe, the United States and Japan. As yet there has been very limited application of this kind of system in the United Kingdom. To the extent that there is investment in new plant for welding in the high-volume areas of engineering production in the United Kingdom, we would expect these types of systems to constitute a significant proportion of that investment.

Applications of this type of system (robot operation plus mechanized handling) in arc welding are less widespread even in the automobile industry than are resistance welding applications, and we would expect whatever application there is in the United Kingdom to be heavily concentrated in the automobile industry, with a particular focus on higher-volume operations in the commerical vehicle sub-sector. Neither the area of application, nor the rate of application within it, are likely to be significantly altered by developments in electronics technology alone.

3. Systems which Incorporate with Robot-type Welding Operation More Advanced Forms of 'Adaptive Control' than Those Currently Available – with or without Mechanised Handling Systems

For resistance welding there are already available adaptive control devices which make intensive use of electronics technology to adjust the welding process parameters to detailed variations in operating conditions (e.g. variations in materials and/or current flows). In high volume applications, which incorporate mechanized handling systems, such devices already seem to justify their costs. The more complex problems of detailed adaptive control of the arc welding process have, as yet, limited the use of this type of system.

For both resistance and arc welding, the extent of likely application is facilitated by the fact that, to some extent, it is possible to 'retrofit' such devices to existing systems – at least in the case of high volume systems. The application of adaptive control devices outside the area of high volume production is likely to be more variable than that within it. In resistance welding applications, it is not clear that the benefits of use of lower-volume production will justify the costs involved. In arc welding applications things may be more clear cut.

4. Systems Involving More Complex and Relatively centralized Computer Control over Multi-machine Welding Systems

Considerable diversity exists in the degree of centralization of the electronic control systems for welding operations. At one extreme the individual welding machine is an isolated entity with its own control system. At the other, large numbers of individual machines can be linked to large central computers with only rather simple 'buffer' control systems attached to each machine. To a large extent the use of the latter systems depends on the simultaneous or prior application of robot controlled welding with or without mechanized transfer systems, and with or without adaptive control devices. To the best of our knowledge use of such systems is at present very limited

TABLE 4.4 Summary of current and future use of advanced electronics-based welding systems

<table>
<tr><th></th><th colspan="3">Areas of production</th></tr>
<tr><th>System categories</th><th>High volume – (mass) production</th><th>Large batch and high volume</th><th>Medium and small batch</th></tr>
<tr><td>1. Robot operation</td><td></td><td></td><td></td></tr>
<tr><td>(a) resistance</td><td rowspan="2">Few current applications, future spread into lower-volume applications likely
***</td><td>Few current applications – might spread in a high growth climate
*</td><td>'No' (?) current applications, future applications unlikely</td></tr>
<tr><td>(b) arc</td><td>Few current applications. Future spread very limited (needs (3) below)</td><td>Some applications especially in 'heavy' mechanical area. Some spread, but depends on (3)</td></tr>
<tr><td>2. Robot operation plus integrated mechanized handling</td><td></td><td></td><td></td></tr>
<tr><td>(a) resistance</td><td>Few current applications, future applications very likely (with investment)
***</td><td rowspan="2">'No' (?) current applications, future spread only with fundamental changes in economic conditions
*</td><td>Currently and prospectively irrelevant</td></tr>
<tr><td>(b) arc</td><td>Very few current applications, future applications likely (with investments)
***</td><td>Currently and prospectively irrelevant</td></tr>
<tr><td>3. Adaptive control with (1) or (2)</td><td></td><td></td><td></td></tr>
<tr><td>(a) resistance</td><td>Some current applications within 2(a), future application increasingly likely within 2(a)
***</td><td>'No' (?) significant current applications within (1). Significant spread unlikely – as for (2)
*</td><td>No significant applications currently or prospectively</td></tr>
</table>

TABLE 4.4 *Continued*

System categories	*Areas of production* *High volume – (mass) production*	*Large batch and high volume*	*Medium and small batch*
(b) arc	Few significant current applications, future spread likely within 2(b) **	No current applications. Significant spread likely following further technical development **	
4. *Centralized computer control of (1) or (2) ±3*	Very few current applications, future applications possible **	'No' (?) current applications, future applications not very likely	No current applications, any significant future applications unlikely

in the United Kingdom – even in the high-volume areas of production in or related to the automobile industry. Their use appears to be more frequent in other countries.

Table 4.4 sums up our impressions of the current and likely rate of application of the different types of system in three broadly differentiated segments of the industry over the next five to eight years.

The following symbols are used in the Table:

*** Some change has occurred and some change will occur.
** No change has occurred, but some change will occur.
* No change has occurred, some may occur (if basic economic conditions change).
No change has occurred. No change is likely to occur.

From the distribution of these symbols in the Table one can see that:

(i) various types of change are likely as a result of the introduction of new types of welding system into the mass production segments of the industry;
(ii) across medium, small batch and unit production, change is likely to occur with the introduction of robot-controlled systems for arc welding which incorporate improved 'adaptive' control devices;

(iii) with rather fundamental change in economic conditions, some of the systems which, under existing kinds of conditions, are largely confined to high volume production, might become applicable in segments of the industry producing in large batches and at the lower end of the high volume sectors.

EMPLOYMENT TRENDS AND SKILL DISTRIBUTION

The number of skilled welders in the industry has remained roughly constant (through fluctuations) from 1969–1977 – at a level of about 45,000–49,000 (see Table 4.5). The number of all skilled craftsmen in the industry has fallen during this period, as has total employment. In consequence, skilled welders increased as a proportion both of skilled craftsmen (from about 5 per cent to 6.5 per cent) and of all employees (from about 1.3 per cent to about 1.6 per cent). These trends are shown in Figure 4.1.

TABLE 4.5 Employment of skilled welders, craftsmen and total employment in the engineering industry

	1969	*1973*	*1977*
Skilled welders	45,450	45,790	48,670
All skilled craftsmen	931,510	747,240	765,400
Total employment	3,492,610	3,110,100	3,127,850

Source: Department of Employment Gazette.

Despite the apparent constancy in the absolute level of demand for skilled welders, and rising relative demand (relative e.g. to other skilled craft groups), data on unemployment and unfilled vacancies for the period 1973–78 shows that the skilled welder category consistently had the lowest or second lowest ratio of unfilled vacancies to registered unemployed.

This larger picture hides some significant differences between sectors within the industry. One sector alone (Industrial Plant and Steelwork) accounts for about 25 per cent of the total employment of skilled welders (see Table 4.6). Six sectoral groups (accounting for about 57 per cent of total employment)

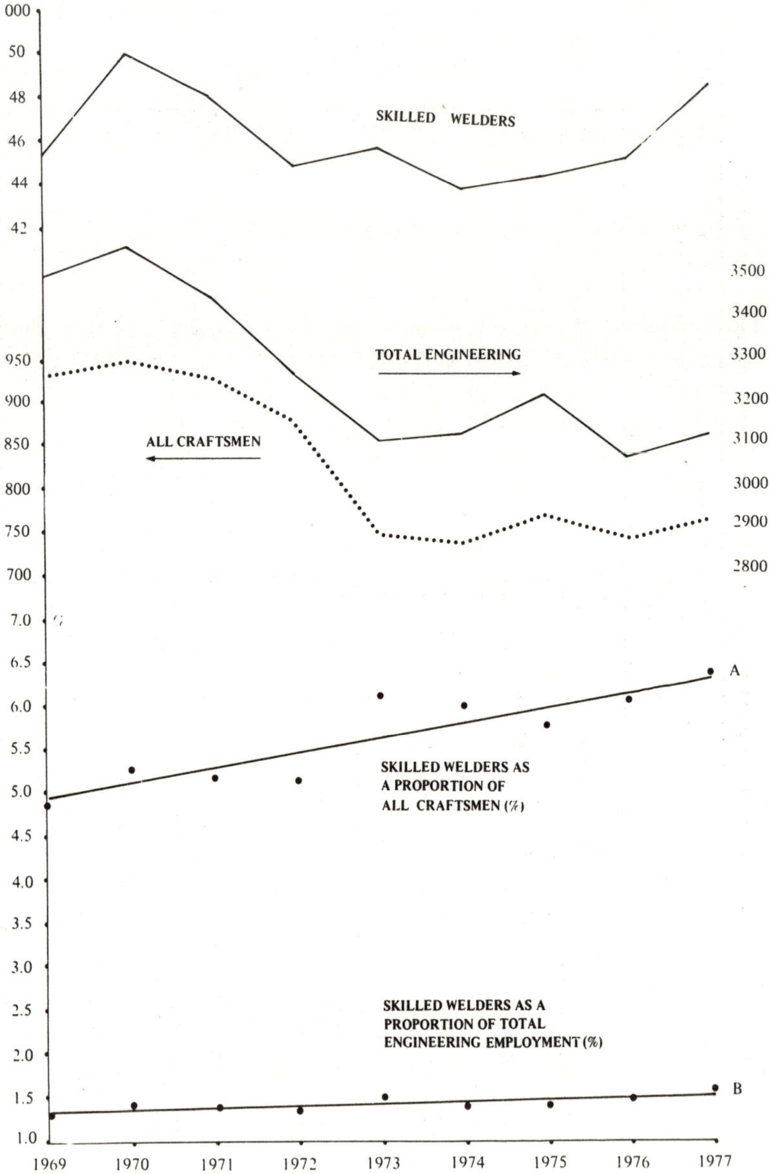

Fig. 4.1 Employment of Skilled Welders, Craftsmen and Total Employees in Engineering and Related Industries, 1969–1977

TABLE 4.6 The intersectoral distribution of welding equipment, skilled welder employment and total employment

MLH	*Sector*	*The distribution of welding equipment 1976* %	*The distribution of total employment 1977* %	*The distribution of skilled welder employment, 1977* %
380/81	Tractors and motor vehicles	19.3	16.9	14.4
333,338/39	Pumps + office and other machinery	17.3	10.6	11.9
399	Metal industries nes	16.7	10.5	16.5
341	Industrial plant and steelwork	9.7	5.2	24.7
363/64/65/66/67	Radio, telephone & electronic equipment	8.6	12.7	0.9
394	Wire and wire manufacture	4.3	1.3	0.4
		75.9	57.2	68.8
361	Electrical machinery	3.8	4.5	3.0
349	Other mechanical engineering nes	3.2	5.7	7.2
369, 362	Other electrical goods, wires & cables	2.6	5.2	0.3
351/53/54	Instruments + photo & copy equipment	2.5	4.5	0.5
337	Mechanical handling equipment	1.8	2.0	5.0
370	Shipbuilding and marine engineering	1.6		
383	Aerospace equipment	1.5	6.3	2.4
368	Domestic electric appliances	1.3	2.1	0.6
336	Construction equipment	1.2	1.4	4.6
331	Agricultural machinery	1.0	1.0	2.1
335	Textile machinery	0.9	0.8	0.6
390	Engineers' small tools & gauges	0.9	2.0	0.3
384/85	Railway locos. & rolling stock	0.6	1.4	3.0
391	Hand tools and implements	0.4	0.6	0.3
332	Metal-working machine tools	0.3	2.1	0.5
395	Cans and metal boxes	0.2	1.0	0.1
393	Bolts, nuts, screws, etc.	0.2	1.1	0.2
334	Industrial engines	0.1	1.0	0.4
		100.0	100.0	100.0

TABLE 4.7 Differences in the 'skill intensity' of welding between the more 'welding intensive' sectoral groups

Group	Code	Sector	*Welding machinery as a proportion of all metal-working equipment* %	*Skilled welders as a proportion of total employment* %	*Skilled welders employed per welding machine used* %
	341	Industrial plant and steelwork	10.7	7.4	2.1
	331	Agricultural machinery	7.8	3.3	1.8
A	336	Construction equipment	7.6	5.0	2.7
	384/85	Railway locos. & rolling stock	7.3	3.3	3.7
	337	Mechanical handling equipment	5.7	3.9	1.6
		Total: Group A	9.0	5.6	2.1
	394	Wire and wire manufacture	20.3	0.6	0.1
	363/64/65/66/67	Radio, telephone & electronic equipment	7.4	0.1	0.07
B	333, 338/39	Pumps + office and other machinery	6.7	1.8	0.5
	399	Metal industries nes	6.7	2.5	0.7
	361	Electrical machinery	5.4	1.0	0.7
		Total: Group B	7.1	1.3	0.5
C	380/81	Tractors and motor vehicles	8.5	1.3	0.4
	368	Domestic electric appliances	5.7	0.4	0.2
		Total: Group C	8.3	1.2	0.4

account for about 76 per cent and 69 per cent respectively of all welding equipment and of all skilled welder employment.

Even within the more 'welding intensive' sectoral groupings, there is important diversity (see Table 4.7). In the first group, the ratio of skilled welders employed to (enumerated) welding equipment used is about five times as large as it is in the third group. The characteristics of Group A are important. Welding in this group of sectors is relatively skill-intensive and heavily concentrated in fusion welding. At the same time, welding is carried out in this group as part of 'heavy', 'low-volume' engineering production. This group accounts for about 40 per cent of all skilled welders, but only for about 14 per cent of all welding machinery.

Differences between sectors (hiding differences between firms) are important for a number of reasons:

(i) Any policies designed to influence technical change relating to welding, and/or the associated skill requirements, can be focused on quite a limited part of the engineering industry while covering a substantial part of total welding activity.

(ii) The aggregate demand for welding-related skills in the industry is likely to be heavily influenced by the sectoral composition of industrial growth.

(iii) The nature of electronics-based 'automation' will differ between sectors (and firms), with different types of production, as will the implications of such 'automation' for employment:

> the types of 'automated' welding system which are applicable will vary between sectors, and with that the demand for different kinds of skills;
>
> the rate at which applicable types of system are actually introduced will vary with these kinds of difference between the sectors;
>
> with a given type of automation and given rate of its introduction, there will be differences in the implications for change in employment, i.e. for the difference between the 'new' levels and structures of skills demanded

and the 'existing' levels and structures. The latter may vary even if the former are 'given' by the type of automation and its rate of introduction.

FUTURE SKILL DEMANDS

The introduction of new types of welding system is likely to be heavily concentrated in (a) the high-volume area of production – especially in those areas in and around the motor industry, and in (b) the 'heavy' end of mechanical, structural and, to a lesser extent, electrical engineering.

In the high-volume area of production, there is likely to be a reduction in demand which is likely to be heavily concentrated on the less-skilled end of the skill spectrum. New systems will use machinery to replace the less-skilled labour jobs involved in (a) the relatively simple tasks of manipulating welding devices through repetitive, standard routines – involving in some cases quite routine stop-start and basic monitoring functions, and in (b) the ancillary handling activities which have not yet been mechanized, but which are likely to be so.

In the 'heavy' engineering area, involving bulky items, relatively long weld-cycle times and a heavy concentration on fusion welding (especially arc welding), the reduction in skill demand is likely to be more heavily concentrated in the skilled-welder area. Various forms of welding systems operated and controlled through robot-type devices are likely to reduce the numbers of skilled welders currently used intensively in this type of production.

There will, of course, also be increases in the demand for certain skills. There are likely to be of the following general types:

(i) A demand for additional components of skill and knowledge on the part of skilled welders engaged in programming individual robot-operated welding machines in the 'heavy' area of production. Additional skills and knowledge are likely to be needed in relation, for example, to fault detection and diagnosis, the rudiments of electronic control system technology and detailed work planning.

(ii) A demand for increased electronics-related skills and knowledge within the maintenance area. This may take either or both of two forms:

a demand for an increased number of people with specialist skills in the maintenance of electronic systems;

a demand for an increased component of electronics-related skill and knowledge within the bundle of skills of individuals involved in maintenance work.

These kinds of demand are likely to arise in both the 'high-volume' and 'heavy' areas of engineering production in which electronics-based welding systems are likely to be used.

(iii) A demand for the skills needed to translate detailed production plans into the form needed as input for 'programmable' welding systems.

(iv) A demand for computer operating skills. This demand will arise from the (probably rather limited) application of the more centralized modes of welding system control.

Equipment suppliers usually have available various kinds of training service to cover many of the kinds of skill needs outlined above. User firms themselves often have their own programmes for providing training in these as well. With respect to the increased demands for electronics-related skills and capabilities in the maintenance area, a number of firms (both users and suppliers of equipment – but more often the latter) have expressed various concerns about the 'shortage' of adequate electronics-related maintenance skills.

CONCLUSIONS

(1) Developments in electronics technology are likely to lead to quite substantial differences in the types of equipment available for welding in the British engineering industry during the 1980s.

(2) However dramatic may be the developments in electronics technology *per se*, they are not likely to cause a very significant difference in the rate of investment in welding

equipment. To a large extent electronics-intensive welding systems will be used as a result of new investment in welding equipment. That investment will be very largely a function of all the things which affect investment in general, and will only be slightly more attractive than would otherwise have been the case without microelectronics.

(3) Application of the new systems will probably be relatively concentrated in particular system-type/production-type combinations. The main areas of concentration are likely to be:
   (i) robot-controlled arc and resistance welding systems with or without adaptive control devices and sometimes with centralized direct computer control, but without integrated mechanization of handling in lower-volume applications within mass production situations – mainly in production in and around the automobile sector and possibly in areas such as domestic appliance production.
   (ii) robot-controlled arc and resistance welding systems with increasing use of adaptive control devices, with associated mechanized handling systems and sometimes with centralized, direct computer control. These will be used within the higher-volume segments of the mass production sectors of the industry, as above.
   (iii) robot-controlled welding by individual welding systems and mainly concerned with arc welding. These kinds of system are likely to be heavily concentrated (but not exclusively used) within what we have called the 'relatively "heavy" end' of the engineering industry involved in low volume production in small batches and individual units.

(4) With the possible exception of electronics-related skills for maintenance, we do not believe that these changes will give rise to significant problems about the shortage of skills for using and operating these systems.

REFERENCE

C.M. Huggett and R.M. Bell, 'Microelectronics, Welding and Skills: A review of some trends in the engineering industry', Science Policy Research Unit, University of Sussex, mimeo, September 1979.

PART II INDUSTRIAL SECTORS

A. Vehicles and Mechanical Engineering

It is difficult to assess the impact of any new technology on a sector of industry, without considering the competitive strengths of that sector in the face of international competition. Each chapter in Parts IIA and B therefore contains sections on the industry's historical background, its structure, the competitiveness of British industry and factors affecting competitiveness in the sector under consideration. This analysis provides the context in which the implications of microelectronics must be assessed. Each chapter contains an account of the impact of microelectronics on the sector and an assessment of the resulting principal skill implications.

This Part includes a chapter on the Motor Vehicle Industry and five chapters on sectors of the Mechanial Engineering Industry, which reflect its diversity.

5. The Motor Industry

P. Senker

The aim of this Chapter is to identify the effects of the advent of microelectronics on the problems of the British motor industry as a basis for assessing the new skill requirements that are arising. The main factors in the industry's poor performance are clear, although controversy rages over the relative importance of each factor and the causal relationships between them. The problems identified by the CPRS (1975) still exist for the most part. Indeed, several of the problems have got worse. So their analysis forms a good starting point, but other more recent sources are also used.

HISTORICAL BACKGROUND

The petrol engine was applied to vehicles in 1878 by a German engineer, and Daimler began to make motor cars in Germany in that year. The first cars were made in Britain in 1895. The United States led in mass production very early on – in 1912, American production had already reached 500,000 – twenty times British production in that year (Hays, 1972). At the end of the First World War, the car industry in Britain was very small. In 1925, protective duties were reintroduced and British output, mainly of small cars, began to grow. By 1937, Britain was producing 380,000 cars a year. Car production resumed after the Second World War and rose rapidly to over 1.8 million in 1964.

INDUSTRY STRUCTURE

The median British motor industry plant employed 2,300 in 1970. The corresponding German plant was over three times as large with 7,600 employees (Jones and Prais, 1978). Jones and Prais suggest that large plant size is as much a symptom of success as its cause. Jones and Prais found that the present substantial differences in plant size between Britain, America and Germany are very much the result of developments in the period 1960–75. During this period, American and German plant sizes at the median and upper quartiles increased their employment between a third and a half, while British plants fell in size slightly. They conclude that 'despite the efforts and intention involved in the great British car mergers of the 1950s and 1960s . . . the net result has had little effect in bringing about the larger plant sizes that are typical of other countries.'

As much as half of the total industry's employment in both Britain and America is in component-making plants. But in Germany, integrated car plants are more usual. Jones and Prais found no overwhelming advantage in the German pattern of large integrated plants, as opposed to the British and American pattern of specialized plants.

COMPETITIVENESS OF THE BRITISH INDUSTRY

The British car industry's share of total Western European and Japanese production declined steadily from nearly 27 per cent in 1963 to under 11 per cent in 1973 (CPRS, 1975, p.59). Between 1965 and 1975, the share of imports in the British domestic market rose from 5 per cent to 35 per cent. In 1979, the import share had risen to 56 per cent compared with 49 per cent in 1978 (SMMT).

FACTORS AFFECTING COMPETITIVENESS

There are six major factors affecting the competitiveness of the motor industry outlined here, though their relative importance remains debatable.

1. *Products and Product Range*

At the time of the CPRS report the fuel crisis had expanded demand in the mini-car segment of the market, but British products were already outnumbered. British Leyland, though they had originated this market, had no new models. But Ford was about to introduce their Fiesta. Some British models were regarded as dated by comparison with their overseas competitors. The Mini and Chrysler Imp were already over twelve years old. In the medium car segment, the Cortina had been launched in 1970, since then a further model had been introduced. (The BL contenders in this sector, the Allegro, Marina and Triumph Dolomite, are considered out of date.) In the executive segment of the market, the Triumph 2000 and Rover 2200 series were also at a disadvantage in relation to their principal European competitors but this was expected to be alleviated by the introduction of the new Rover range.

By the end of 1974, the price advantage of British cars had already disappeared. British cars had acquired a poor reputation for quality and this had already adversely affected dealers' attitudes to British made cars both at home and abroad (CPRS, 1975, pp. 64–75).

2. *Labour Productivity and Capital Expenditure*

For the CPRS, the principal cause for concern was low labour productivity in the British car industry, particularly in car assembly plants: productivity in engine and transmission production was found to be below continental levels, but not to the same extent (CPRS, 1975, pp. 80–1). Subsequently, Jones and Prais found that British productivity in car manufacture was slightly higher than Germany's in 1955, with American productivity nearly five times higher. By 1965, German productivity was a tenth higher than in Britain and by 1976, German productivity was 40 per cent higher than in Britain (Jones and Prais, 1978).

For several reasons, even where production lines were manned at competitive standards, output in British plants was found to be relatively low. The causes identified were:

slower line speeds, late starts, frequent stoppages between and during shifts and delays in correcting mechanical problems;

shortage of materials due to labour disputes affecting suppliers or to poor internal management (CPRS, 1975, pp. 82–4).

The significance of capital expenditure in competitiveness remains a matter for debate. While the CPRS Report argued that 'inadequate capital expenditure is only a minor cause of lower productivity', the House of Commons Expenditure Committee concluded that 'until investment substantially increases the capital stock, productivity in the motor industry will not be able to equal productivity abroad' (CPRS, 1975; Expenditure Committee, 1975).

3. *Overmanning*

The CPRS considered that overmanning, e.g. in assembly and maintenance, was a principal cause of poor productivity. The example was given of a multiweld machine, where because of trade demarcation, six maintenance men of different skills would be required to repair it, while on the continent, two men only could accomplish the same job. This type of consideration is particularly relevant to the manning requirements for maintenance of machines embodying substantial microelectronics content – e.g. robot welders.

4. *Interruptions to production*

The high degree of integration of the motor industry, the high volume of production in relation to low levels of buffer stocks, and the need to halt assembly lines whenever any one component cannot be applied, or whenever any station is not operating, render the industry highly sensitive to interruptions to production. The extent to which industrial disputes are responsible for these is a matter of continuing debate.

In 1969–72, 8.1 million working days were recorded as lost as a result of disputes. This represented approximately 1½ per cent of the total working days available for production. If 65 per cent is added to Department of Employment figures for days lost to account for lay-offs as a result of strikes occurring within the industry, the loss of time would have represented about 2½ per cent of total working days. But industrial disputes are only one aspect of interruptions to production. Research in 1969 suggested that labour disputes accounted for about one-half of estimated lost production, the rest being due to breakdowns, changed customer schedules, technical change etc. (NEDO, 1973).

The Ryder Report suggested also that British Leyland's stoppages were 'often the result of factors outside the control of the workforce: breakdowns in plant and equipment, faulty scheduling, shortage of materials and components, and external industrial disputes' (EITB, 1978).

5. *Quality*

The high incidence of quality faults, by increasing the time and and effort required for fault rectification, also has an important effect on productivity. Despite the fact that British manufacturers employed 50–70 per cent more plant maintenance personnel than continental competitors on identical equipment, mechanical breakdowns resulted in the loss of about twice as many production hours in the United Kingdom (CPRS, 1975, p. 84).

6. *The Industry's Fragmented Structure*

The fragmented structure of the British motor industry results in considerable duplication of effort, leading to increased operating overheads which are a very important part of total costs. Poor labour productivity and production losses increase the cost of capital for each car produced. Because car plants in Britain consistently fail to achieve the same output from capital equipment comparable or identical with that on the continent, the cost of the equipment has to be depreciated over a smaller number of cars (CPRS, 1975, p. 91).

IMPLICATIONS FOR OTHER SECTORS OF THE ENGINEERING INDUSTRY

The continued decline of the British owned vehicle assembly industry has serious implications for the components sector.

The components industry relies on close relationships with vehicle manufacturers to design and manufacture its products; 'components are not designed and produced in isolation from vehicle manufacture' (Smith, 1979). It is essential for British component manufacturers to be involved in new model developments by working closely with the assemblers originating new designs, not only contributing to research but also taking a share in ownership of tooling and facilities. This becomes more important with model rationalization and the trend towards 'the world car' with basically standard components but sourced from a number of countries (Smith 1979).

The weaknesses of BL as the only major British indigenous car assembler, both in terms of market share and of research and development, are important not only in themselves but also insofar as these weaknesses impinge on the components industry. In 1978, Michael Edwardes admitted to The Select Committee on Science and Technology (316–v) that BL was investing a smaller percentage of turnover in Research and Development than many of its international competitors. But he claimed that the company was more successful in that the quality and imagination of their R & D was such that BL's R & D was more effective per unit of effort (p. 49). At this time, BL were running experimental vehicles incorporating microprocessors and had initiated a development contract for microprocessor work in the United States (p. 68).

But a colleague, D.R.G. Andrews, pointed out that not only is BL spending less on R & D but that competitors' spending is devoted to a narrower product range and that they are producing increasing numbers of vehicles which have more universal appeal. The ability of American-based firms to spread their development over more of their worldwide operations is increasing as the size of American cars is reduced to that more typical of European cars. This gives the large American producers a very keen competitive edge (p. 59).

BL consider that the tremendous growth in technology, particularly in the electronic and electrical field has left all automobile manufacturers short of skilled personnel. BL are particularly short of electronics and electrical engineers and creative design engineers. The shortage is considerable,

TABLE 5.1 Comparison of Ford and BL

	Ford	*BL*
Total employment	67,000	170,000
No. of people earning		
£12,500–15,000	113	58
£10,000–12,500	779	142

but less than the 900 estimated previously. BL attributed this to pay restraint, producing the comparative figures shown in Table 5.1.

BL considered that Ford started with higher salary levels and that BL salaries were pegged by pay restraint after a period when the company was so short of money that salary increases had been limited anyway. Not only is BL unable to attract the people it needs, but it does not invest anywhere near as much in R & D as its major competitors, because its profitability is so low that it has not got the money to back research to the extent that it would like to (Select Committee, 316–v, pp. 49–51).

As BL is the only major British owned vehicle assembler, and considers that its already low R & D expenditure is constrained by its poor profitability, the implications for the component industry appear very serious.

THE IMPACT OF MICROELECTRONICS

The major preoccupation of motor manufacturers throughout the world is the drive to achieve fuel economy. Ever increasing oil prices and pressure from governments to reduce countries' dependence on overseas oil suppliers are likely to reinforce this drive. In the next few years the principal applications of microelectronics to motor cars are likely to be where they can contribute to fuel economy; but in achieving this objective, weight reduction is likely to be more important than microelectronics applications.

The United States is leading in the application of reinforced plastics. Plastics are established in many components – e.g. moulded bumpers. Perhaps the most interesting process under development at present is reaction injection moulding – RIM. Both Ford and General Motors are working on this.

A filament-wound 70 per cent glass fibre/epoxy resin leaf spring may be in production by General Motors in 1981. The spring weighs 8lb. compared with the 45lb. of the multi-leaf steel spring now used, and has shown several times the durability on test. Drive shafts and other components have also been tested in the United States, but there is, as yet, no published information on plans for incorporation in production vehicles. In Britain, GKN and Bristol Composite Materials Engineering have developed carbon-fibre based springs intended initially for commercial vehicles. Application of such plastics components seems further off in Britain than in the United States (McGeehin, 1979).

In the immediate future, monitoring and regulating engine functions may well be most important areas for microelectronics applications. But manufacturers are also working on electronically controlled automatic transmission systems which are planned to be more economic on fuel than manual transmissions. Safety applications and instrumentation are also probable areas of application, as also are emission control and vehicle condition monitoring, but these are likely to be less immediate than fuel economy applications. Initially, solid state instrumentation may be no cheaper than conventional instrumention, but may well be incorporated for 'fashion' reasons. Widespread use of on-board radar and diagnostics on vehicles are unlikely in the immediate future for cost reasons. The conditions under the bonnet of vehicles are very demanding for electronic components. Temperature and humidity vary widely. Electrical noise can also be a problem and the actuators used to convert electrical signals into movements of one kind or another can contribute to noise (Burkitt, 1978).

McLean and Rush (1978) suggested that, in the early 1970s, the major motor vehicle manufacturers in the United States held back from using microprocessors for monitoring and regulating engine functions because they expected enormous improvements in microelectronics. But subsequently, simultaneous developments in semiconductor technology and energy, environmental and safety legislation forced the American companies to take microprocessors more seriously.

E.R. Karrer, from US Ford, has been reported as describing

the 'control of the automotive powertrain' – the control of the whole process by which power is created in the engine and converted into motion – as the most crucial area of application for electronics in the car industry. The three major American manufacturers have been introducing frequent changes to their electronic engine and transmission control modules. Each design is significantly smaller, less costly, more powerful and faster than its predecessor. Electronic modules are only part of the problem, as modules need appropriate sensors for temperature, pressure, crankshaft position and exhaust composition (Boothroyd, 1979).

It is particularly significant that microcomputers have been applied to carburettors as well as fuel injection systems, as carburettors are cheaper than fuel injection and are used in the popular mass produced cars. In August 1979, it was reported that Ford are to use microcomputers 'in a new feedback carburettor system designed for Ford's four cylinder cars such as Mustangs and Pintos. The system is designed to minimize exhaust emissions while boosting fuel economy' (Mitchell, 1979). The microcomputer receives information such as engine speed, oil pressure and temperature, water temperature and atmospheric pressure. Then it calculates to produce signals which control constantly changing elements such as air–fuel ratios, exhaust gas recirculation rate and the optimum time in the cycle for the spark plug to fire. Mitchell comments that the Ford contract with Intel to supply the microcomputers is evidence that the industry is going to rely on electronics increasingly to meet stringent federal emission and fuel economy standards. Similar technical developments are known to be going on in the United Kingdom, but no details are yet available. The main advantages are fuel savings. It has been suggested that electronically controlled ignition, emission gas recycling and air–fuel mixing rate can reduce a car's fuel consumption by at least 10 per cent (McIntosh, 1979).

In 1978, General Motors said that they saw 'a significant use of microprocessors in fuel and ignition controls particularly' (Select Committee, 263–iii, p. 112). In 1979, the AC Delco Division of General Motors agreed to buy five million microprocessors and also memory parts from Motorola and Fairchild for a total of about £20 million spread over two years. Ford is also increasing its microprocessor purchases very substantially buying from Toshiba, Motorola, Texas Instruments and Intel (McIntosh, 1979).

By 1978, Chrysler Detroit had developed a 'very good le burn system which depends on microelectronics and smaller computers. They have a closed loop system which will sense exhaust emissions and correct fuel input etc.' (Select Committee, 263–ii, p.101). Chrysler UK claim to have been the first major manufacturer to introduce electronic ignition in Europe. The basic reason for putting it on the Sunbeam 930 was that in their original design, the conventional distributor was inaccessible for setting and maintenance. Although electronic ignition was more expensive, it was justified because, not needing to be set, the inaccessibility of the electronic ignition was not so important (Select Committee, 263–ii, pp. 112–3).

As McLean and Rush point out, whilst the majority of component manufacturers in the United States are subsidiaries of American automobile firms, this is not so in the United Kingdom. Companies independent of the automobile manufacturers, such as Smiths and Lucas, are important as exporters and innovators (McLean and Rush, 1978, p. 35). For example, Lucas have developed both microprocessor based fuel systems and microprocessor controlled ignition. Their fuel system is a digital petrol injection system for cars, for which they claim petrol savings of between 5 and 25 per cent, achieved by matching fuel requirements to different engine speeds and loads more closely than the analogue system (*Electronics Times*, 1979).

SKILL IMPLICATIONS

The motor industry has a great need for electronics engineers to design systems for controlling fuelling, ignition, transmission and instrumentation, for safety applications – for example for brake control, and for controlling the distribution of electricity round the vehicle. Electronics engineers are also needed for research on projects such as on-board radar and diagnostics.

The general need is for electronics engineers specializing in control theory who understand the applications of micoprocessors, and who also understand the principles of engines and transmissions. The basic need is for electronics engineers who can use electronics to control mechanisms. Although motor car and component firms may have difficulty in matching the salaries offered by electronics firms, a more fundamental problem is that few electronics engineers want to work on projects where mechanical considerations are important. The design of transducers-sensors and actuators also requires graduates who need a knowledge of physics and mechanical engineering with some electronics. It is not quite so difficult for motor car companies to recruit such people.

REFERENCES

David Boothroyd, 'Micro prospects seem set to be très bon under the bonnet', *Electronics Times*, 15 November 1979.

Alan Burkitt, 'Bleak outlook for microprocessors in the motor industry', *The Engineer*, 23 March 1978.

Central Policy Review Staff, *The Future of the British Car Industry*, HMSO, London, 1975.

Electronics Times, 'First digital car system is from UK', 12 April 1979.

Engineering Industry Training Board, *Employment and training in the motor vehicle industry*, EITB Reference Paper 2, 1978.

Samuel Hays, *The Engineering Industries*, London, Heinemann, 1972.

House of Commons Expenditure Committee, *The Motor Vehicle Industry*, 1975.

D.T. Jones and S.J. Prais, 'Plant-size and productivity in the motor industry: some international comparisons', *Oxford Bulletin of Economics and Statistics*, May 1978.

Peter McGeehin, 'The Plastic Car: why Europe should follow in America's tracks', *The Engineer*, 15 March 1979.

T. McIntosh, 'Ford doubles its order for electronic controls', *Electronics Times*, 12 April 1979.

J.M. McLean and H.J. Rush, *The Impact of Microelectronics on the UK: A Suggested Classification and Illustrative Case Studies*, SPRU Occasional Paper Series No. 7, June 1978.

James Mitchell, 'Ford to use micros in 1980 models', *Electronics Times*, 2/9 August 1979.

National Economic Development Office, *Motors* (industrial review to 1977), 1973.

The Select Committee on Science and Technology, Technological Innovation Sub-Committee, *Developments in Small Passenger Car Engine Technology*, Session 1977–78, 316–v, HMSO, London, 1978.

The Select Committee on Science and Technology, Technological Innovation Sub-Committee, *Developments in Small Passenger Car Engine Technology*, Session 1978–79, 263–iii, HMSO, London, 1978.

The Select Committee on Science and Technology, Technological Innovation Sub-Committee, *Developments in Small Passenger Car Engine Technology*, Session 1978–79, 263–ii, HMSO, London, 1979.

A. Smith, 'Motor component manufacturers in need of a home market', *Financial Times*, 9 February 1979.

Society of Motor Manufacturers and Traders.

6. Forklift Trucks

P. Senker

Forklift trucks are a particular class of mechanical handling machine. They are power operated, mobile lifting units, equipped with forks, adjustably attached to a carriage running in a mast, for the lifting, transporting and stacking of unit or palletized loads. Forklift trucks operate along 'free paths'. They are in competition with other forms of materials handling devices – in particular with 'fixed path' devices such as conveyors and stacking cranes.

Forklift trucks are 'powered mobiles' (as also, for example, are tractors and excavators), assembled mainly from parts also used by the automotive industries (e.g. transmission components), and partly from more specialized components (e.g. hydraulics, forks, fork carriages). Bought-in components account for a substantial proportion of works-costs for all forklift truck manufacturers – as high as 85 per cent in some cases.

Because of the relatively low proportion of value added accounted for by manufacturing activities, forklift truck manufacturers deploy a high proportion of their manpower in such service functions as purchasing, stock control, kitting up, design, sales, marketing and after-sales service (including provision of spare parts).

HISTORICAL BACKGROUND

The first *powered* industrial trucks were introduced in the United States in 1906 and a few years later in the United Kingdom.

The First World War gave impetus to the design, development and wider application of several types of industrial truck. The first counterbalanced forklift trucks were introduced in 1917, but their use developed slowly. The great stimulus to the wider use of counterbalanced forklift trucks was provided by the Second World War by the American armed forces. This was copied by the British armed forces. Forklift trucks then spread to British industry. A British firm produced its first (counterbalanced) forklift truck in 1947.

The British forklift truck industry has grown rapidly in the last twenty-five years. The industry started with the use of general purpose counterbalanced forklift trucks in an increasingly wide variety of applications. Whilst the general purpose truck still serves adequately as a mechanical handling workhorse, it has some disadvantages for some applications. Manufacturers have developed a large number of variants of forklift trucks, in order to mechanize a wider range of materials handling operations. Various sources of motive power have been used – principally internal combustion engines for outdoors and electric motors for indoor use. General purpose counterbalanced trucks can be used for warehousing applications, but they require comparatively wide aisles in which to operate. In addition, the height to which they can stack goods is limited. Particularly in Europe and Japan, where space costs are high, such limitations have led to the development of varieties of narrow-aisle trucks, which can operate in confined spaces and stack higher than conventional counterbalanced trucks.

The development of containerization has led to the development of very large counterbalanced forklift trucks and large sideloading trucks, as well as machines based on different principles altogether (straddle carriers). Compared to manufacturing industry, construction and agriculture have been relatively late in starting to mechanize materials handling. To meet these needs for mechanized handling in difficult conditions, the rough terrain forklift truck was developed.

In parallel with the development of a greater variety of types, there has been steady development of performance arising from the adoption of improved components. For example, electronic controls, principally based on thyristors, have been adopted widely on electric trucks. Compared with the conventional electromechanical controls, these reduce battery drain, enabling electric trucks to operate longer or more arduous shifts. Microelectronics makes possible the development of more sophisticated control systems.

INDUSTRY STRUCTURE

The British forklift truck industry is fairly concentrated. There may be as many as thirty forklift truck manufacturers in the United Kingdom, the majority British-owned. As a result of recent mergers, it is probable that six firms, three of them American-owned multinationals, account for about three-quarters of output. Three firms account for about half the industry's output.

COMPETITIVENESS OF THE BRITISH INDUSTRY

The United Kingdom accounted for about 7 per cent of world production of forklift trucks in 1974 (NEDC, 1977, p. 14). While the United Kingdom has maintained its share of world export markets in the last few years, the value of imports of forklift trucks to the United Kingdom increased very fast, more than doubling as a proportion of home consumption between 1970 and 1977 (Senker *et al.*, 1977).

FACTORS AFFECTING COMPETITIVENESS

Important in the competitiveness of forklift trucks is their efficiency and reliability and the overall costs associated with their operation. These costs include not only the initial purchase price and the costs incurred in operating and keeping trucks in operation (drivers' wages, fuel, maintenance, etc.), but also costs imposed on customers' total materials handling systems by downtime. Inevitably, all forklift trucks, however well designed and constructed, break down at some time and require spares and service. Customers, especially those who use their forklift trucks intensively, have a great interest in ensuring that, when the forklift trucks do break down, service and spares are readily available, so that expensive downtime is minimized.

There have been two basic routes towards success in the forklift truck business. One involves establishing market leads through product innovation and preparedness to modify trucks to meet varying customer needs. The other involves cost reduction through component standardization between different models of trucks and, in the case of multinationals, across countries, in order to reduce production and other costs (Senker *et al.*, 1977, pp. 48–9).

Even though this strategy has been the one favoured by the American multinationals who have been relatively successful, the American firms have not achieved sufficient competitive strength to impose standardization on the market. The American-owned multinationals are planned centrally from American headquarters. They all place strong emphasis on marketing and after-sales service, employing graduate engineers in marketing and sales roles to a far greater extent than British firms do.

The major problem likely to face the British forklift truck sector over the next several years will be the maintenance of its generally good record in incremental product innovation in a changing market and technical environment. Innovation is likely to be based increasingly on formal research and development, and there is likely to be a need for more technical depth in the information feedback firms receive from markets. These trends suggest an increased need for centralized planning and control of all aspects of firms' policies.

THE IMPACT OF MICROELECTRONICS

In a period of slow technological change, an important competitive weapon has been to offer to meet customers' needs more closely by modifying trucks. Technology is now changing more quickly – although this should not be exaggerated – so that those firms which adopt microprocessors more quickly in their products may secure product leads which enhance their market position significantly. Two examples are:

(1) Truck automation. A technical lead in developing efficient, labour-saving systems could create a market lead in the warehousing sector;

(2) Truck controllers. Development of compact, cheap, reliable controllers coupled with automatic fault diagnosis could help establish a market lead through a reputation for reliability and the ability to promise less truck downtime than is possible with current systems.

The problems for the industry, and for the British-owned part of it in particular, is that there are a multiplicity of ways in which microelectronics could improve the efficiency of forklift trucks and of the materials handling systems in which they operate. It requires substantial technical efforts even to evaluate the principal possibilities so as to assess which are likely to be most in demand in the near future. The best return could emerge from fairly radical developments requiring very substantial technical manpower resources to achieve. Brief accounts of some of these possibilities follow. In some areas, systems embodying microelectronics have already been installed.

A. Motor Control

A very wide range of motor controllers is in use on forklift trucks which are in production in the United Kingdom, ranging from electromechanical controllers designed in the 1950s to sophisticated electronic controllers incorporating microelectronic devices.

Electromechanical controllers are still used partly because of the conservatism of the British materials handling equipment market. In small forklift trucks and pallet trucks, power consumption may not be critical to the user or, in some cases, the electromechanical control system used incorporates regenerative braking, tending to improve its power consumption relative to electronic systems. Perhaps more important, users are often anxious to continue to buy the same controllers they have bought for many years to minimize their spares' holdings.

Microelectronics offer the possibilities of reducing the costs of controllers, through reducing the number of components they incorporate. Indeed, some manufacturers have made substantial progress in this – at least one manufacturer, for example, now uses complementary metal oxide semiconductor

(CMOS) integrated circuits in controllers. As well as the traction controller, pump motors, hydraulic motors and the rotating heads on turret trucks now employ relatively sophisticated electronic controllers. Power steering is mainly hydraulic – in future it could be controlled electronically.

B. Fault Diagnosis

Microelectronics can be used to aid fault diagnosis. For example, some faults may be intermittent. By the time the service engineer arrives, the fault has cleared and he has no way of telling what has caused it. The use of microelectronics makes it possible to record fault conditions as they occur.

Maintenance and service are undertaken by forklift truck manufacturers, by users themselves and by specialist service firms. Fault diagnosis at present consists mainly of using specified procedures with a multimeter. It is possible that automatic test equipment will be developed for use by service engineers.

C. Battery Control

The continuous monitoring and control of battery charging is now possible using electronics. A charging system, claimed to be the first to do this, was introduced recently. It was claimed that this system makes possible charging of batteries in any condition and circumstances without supervision (*Materials Handling*, 1979a).

Such systems are likely to become increasingly sophisticated, including monitoring the temperature, age and condition of batteries to feed back information to the controller which controls the rate of charge. The capital cost of batteries is high, the amount of energy a battery can deliver between charges is often important to users, and energy costs are increasing. Systems which improve battery life and economize on energy use are likely to become increasingly attractive to users, especially as microelectronics is creating the possibility of the development of increasingly sophisticated but cheap controllers.

D. Safety

A potential area of microelectronics application is to control forklift truck stability. Another possibility is head-up displays for forklift truck drivers.

E. Truck Automation

Computer controlled transport and handling systems are already in operation. By 1978, the Japanese forklift truck company, Komatsu, had already installed eight such systems in Japan, and had signed a further contract to install one in Sweden. The Komatsu system included a lift truck with a microcomputer, wire guidance and optical and ultrasonic sensors to locate load position (Komatsu, 1978).

Recently, sophisticated systems installed in the United States, with computer controls supplied by Logisticon in the United States, were reported in the press. In one such system, the computer had many functions, including the following:

transmits pick instructions to truck operators and generates move tickets through printers mounted on the trucks;

receives information from truck operators through keyboards mounted on the trucks;

maintains inventory and stock location records;

responds to item availability requests which may be entered through standard touch-tone telephone or CRT/keyboard telephones.

In one installation, order pickers are fitted with wire-guidance, independent of the computerized communications systems, for automated steering in narrow aisles (*Materials Handling*, 1979b).

The eventual possibilities are very considerable, an extreme case is robotic order picking. It is doubtful whether the British forklift truck industry will have the resources of technical manpower to pursue such possibilities.

SKILL IMPLICATIONS

The pervasiveness of the technical change being made possible by microelectronics means that detailed technical information about both components and markets is becoming of more importance to the forklift truck manufacturer. Any links that a forklift truck company may have – e.g. with suppliers of complete materials handling systems, with materials handling consultants, or with motor manufacturers – can be useful in improving understanding of market needs and technical requirements. Even where there are firms within the same group of companies involved in these different aspects of the business, the individual units tend to be organized so that they deal with each other on an 'arm's-length' basis. This may be advantageous in terms of managerial accountability for individual profit centres, but it can be a disadvantage in coming to terms with technical change. In general, the problem of insufficient communication of detailed technical market information to design and development, arising out of the failure to employ sufficient highly qualified and trained people in marketing, market research and sales, still persists.

The most urgent requirement of the forklift truck industry is for resources of technical manpower. The minimum nucleus necessary for a major forklift truck manufacturer would probably be at least two highly qualified and experienced systems designers and electronics engineers at graduate level, plus at least two high level technicians and several lower level technicians, programmers etc. We have not been able to survey the industry to establish the extent to which the manufacturers have built up such design teams, but there is no evidence yet that there are any teams of such size and competence in the British-owned part of the industry. The alternative, which is being adopted by at least one major firm, is to employ outside consultants, partly because the firm was unable to recruit the people it needed. The firm found, just as firms in the motor and television industries have found, that the electronics people it required were anxious to work in the technological frontier areas of the electronics industry. Further, the skills required to interface mechanical and electronic systems are extremely rare – a similar problem to that faced by the motor industry.

Whilst employing consultants can be helpful, it is likely to be insufficient to solve the industry's problems. For example, while the industry may be able to get consultants to develop systems which result in incremental improvements, it is unlikely that radically improved systems – e.g. radically improved motor controllers or systems for truck automation – will emerge from people without a deep understanding of materials handling and forklift trucks themselves. Experience in other industries indicates that the use of consultants, although potentially valuable, is fraught with problems.

REFERENCES

Komatsu Forklift Co. Ltd., News Release, 12 May 1978.

Materials Handling News, 'The power game', July 1979a.

Materials Handling News, 'How a computerised lift truck fleet pays off', August 1979b.

National Economic Development Council, NEDC Industrial Strategy, *Progress Report* of the Industrial Trucks Sector Working Party, April 1977.

P. Senker, E. Sciberras, N. Swords and C. Huggett, 'Forklift Trucks: A Study of a Sector of the UK Engineering Industry', report prepared for the Engineering Industry Training Board; Science Policy Research Unit, mimeo, September 1977.

7. Agricultural Equipment

P. Senker

Microelectronics is likely to be applied less pervasively in agricultural engineering than in some other mechanical engineering sectors. Nevertheless, there are a few areas where successful application of microprocessors to products and systems could give firms a competitive edge. Failure of British firms to apply microprocessors where applications are viable could lead to consolidation of European manufacturers' lead in the production of sophisticated agricultural machinery.

HISTORICAL BACKGROUND

This is a very old industry which has always been scattered through the agricultural regions of the United Kingdom. Between 1885 and the mid-1940s, yields per acre of major British crops changed hardly at all. But after the Second World War, yields rose rapidly (Blaxter, 1973). Principal reasons for the growth in yield have been the application of new machines and new fertilizers. Between 1946 and 1971, the numbers of combine harvesters in use increased from 3,500 to 66,000; grain driers from 1,000 to 63,500 and motor vehicles from 62,000 to 130,000.

Agricultural machinery increased yield in many ways. For example, milking machines increased milk production and freed labour for other tasks; more fertilizer, spread more evenly by machine, increased crop yields; spraying chemicals also increased crop yields by protecting crops from competition from weeds and by preventing the spread of disease by pests.

INDUSTRY STRUCTURE

The industry produces a wide variety of products, from dairy machinery incorporating sophisticated electronics to simple implements which have remained unchanged for a long time. Firms range from large multinationals to small workshops. Most indigenous companies are independent and have been established for a long time, although in recent years many smaller firms have been acquired by large groups. The 1973 Census of Production showed the largest six companies accounting for 40 per cent of industry sales.

COMPETITIVENESS OF THE BRITISH INDUSTRY

In 1966 the value of British exports was over twice that of imports. Between 1966 and 1976, British industry's sales increased by about 6 per cent while the British market expanded by over 30 per cent in real terms. The industry increased its exports by about 30 per cent, but imports trebled to the point where imports and exports were in balance (see Table 7.1).

TABLE 7.1 UK home market, industry sales, exports and imports

	(£m at 1970 prices)			
	Home market	*Industry sales*	*Exports*	*Imports*
1966	70	88	32	14
1976	93	93	42	42

Source: *The Agricultural Engineering Industry*, Report of a Department of Industry Study, London, 1978.

FACTORS AFFECTING COMPETITIVENESS

1. Products and Markets

The production of novel machinery, offering an advantage to the farmer, has enabled relatively small firms to dominate particular market segments. Perhaps the only really radical change in agricultural machinery since the war has been the use of sophisticated electronics in some recently developed milking and feeding systems. Dairy systems have become increasingly automated and productive and, via the use of sophisticated electronics, can be linked to cow identification and feeding systems.

The largest companies operating in dairy machinery are Alfa-Laval (Swedish owned, £10 million turnover in 1973), who make a high percentage of British milking machinery based on Swedish designs; Fullwood and Bland (£8 million turnover in 1973) and Gascoigne, Gush and Dent (£5 million).

In the 1970s there has been a fairly dramatic increase in imports to the United Kingdom of highly priced, sophisticated machinery. Farmers consider foreign machinery to be more reliable, more efficient, better designed and more up to date than British.

2. Production Processes

European manufacturers also emphasize organization of the production sequence. They tended to patent more production processes than British firms. Some are moving towards continuous flow in place of batch production for their major product lines. They were attempting to accommodate the maximum product variety within continuous flow lines. Design and production are better integrated, leading both to better products and to lower manufacturing costs. A number are attempting to improve quality via automation and the use of numerically controlled machine tools which cut down the possibilities for human error.

3. Management and Technical Expertise

The major problem in the United Kingdom appears to be that few firms, especially the smaller firms who comprise most of the industry, have the management expertise to achieve satisfactory long-term planning. Many also lack the technical resources to draw ahead of their larger continental rivals even if they had plans to do so.

Several firms stated that the National Institute for Agricultural Engineering's electronics department provides an excellent service. Two firms which both now possess an electronics capability in-house stated that without the assistance of NIAE's electronics engineers, they would almost certainly have lost technical leads. However, Rothwell (1979) comments that the electronics assembly departments of both these firms have been sited poorly, at the centre of dirty, oily, dust-filled machine shops.

Several small firms in the British industry stated quite explicitly that they did not wish to grow. According to one managing director, control becomes more difficult with increasing size and contact is lost with employees. This man established a separate small firm in a development area with government aid to exploit a new idea, rather than expand his existing firm. Similarly, a second managing director saw his firm as a balanced social unit. His firm is both innovative and profitable but has grown little in the last thirty years. When he developed a new machine requiring large-scale production facilities, he licensed a much larger firm to manufacture and sell it. Both these firms marketed most of their output within the United Kingdom and neither was much interested in exports.

Some farmers buy British because of the convenience of ready local availability of spare parts and because locally built machines are sometimes considered to be more suitable for local conditions. The prices of British machines and spare parts are lower but several British firms showed 'a sad lack of awareness of the importance of quality control' (Rothwell, 1979).

THE IMPACT OF MICROELECTRONICS

Potentially important areas for the application of microelectronics include dairy parlour engineering; livestock condition monitoring – including weighing; planting and spraying monitoring and control; greenhouse environmental and nutrient control, and eventually automated ploughing.

1. Dairy Parlour Engineering

The National Institute for Agricultural Engineering is developing a scheme for automatic control and management in a dairy parlour. The system will identify the cows, record concentrates dispensed, body-weight, milk yield, etc. for each cow. The system can store and present data and reports, which also include clinical and fertility records (NIAE, 1977–8).

This scheme includes four microprocessors distributed about the system to ensure flexibility. With the help of the NRDC, a manufacturer is planning to supply it for initial use on a few experimental farms, hoping to exploit it commercially in the 1980s.

2. Livestock Condition Monitoring

Several small firms already offer pig weighers. The NIAE is developing a system for weighing chicks in batteries. This system could be extended to monitor other variables, e.g. feeds.

3. Planting and Spraying Monitoring and Control

Plant monitoring equipment can ensure that crops are planted evenly, with a minimum of waste of seed. For example, if planting is not monitored, a seed outlet may become blocked and the crop will be planted inefficiently.

4. Greenhouse Environmental and Nutrient Control

Microelectronics makes it possible to devise more sophisticated and flexible control systems for greenhouses. Heating and ventilation can be controlled more closely. In the early stages of growing, a drop in temperature below a critical level can affect crop yields seriously. Thus, closer control of temperature can maximize yield, at the same time keeping energy consumption under close control.

Nutrient film hydroponic systems are being used increasingly to grow such important crops as tomatoes and cucumbers. These systems make possible substantial improvements in crop yields. Increases in yield of over 25 per cent may be possible by the use of nutrient film techniques, using conventional control techniques. However, nutrient film techniques may involve the use of fourteen constituents and conventional control techniques only involve control of pH and conductivity. The use of microprocessor-based control systems could lead to closer control of more variables, leading to further significant improvements in crop yields.

5. Automated Tractors and Combines

Ploughing involves long hours at high tractor power and consequently high stress. The NIAE has developed an automatic guidance method in which the first furrow is ploughed under manual guidance. The positions of this and subsequent furrows are sensed optically and the tractor steered accordingly. Field boundaries are detected using near infra-red light and the signals are used to initiate and control headland turns. Microprocessors are used for signal processing to give improved accuracy of range measuring (NIAE, 1977–78, pp. 14–15). Such systems offer the possibility of a worker supervising and operating more than one tractor and of ploughing at night.

However, development beyond the experimental stage will need to await the surmounting of some very considerable barriers. To be diffused widely, it would need to be adopted by a major tractor manufacturer. At present, tractors generally have manual transmissions. To be able to use the full range of ploughing speeds, a remotely controlled tractor would need an infinitely variable speed drive – e.g. a hydrostatic transmission – and this would add even more to the complexity and cost of the tractor. There was a considerable amount of development effort put into hydrostatic transmissions in the 1960s, but high cost and low power efficiency have militated against them. Only one firm in the United Kingdom makes hydrostatic transmissions and for only one tractor in its range (Coombs *et al.*, 1978, p. 2.8).

Agriculture has had a poor record in relation to health and safety; a significant proportion of accidents have been caused by tractors overturning (Coombs *et al.*, 1978, p. 8.12). This led first to legislation for the introduction of roll bars and subsequently to cabs, but even now cabs are only fitted to a modest proportion of output. In the United Kingdom, the NIAE and the Ministry of Agriculture Safety Inspectorate were prime movers in instigating legislation, and the Ministry co-operated with the National Union of Agricultural Workers in pressing for legislation, against the opposition of both the tractor manufacturers and the National Farmers Union (Coombs *et al.*, 1978, p. 8.14).

By contrast, automatically guided tractors are likely to be opposed by the agricultural workers, and the farmers are not likely to be enthusiastic unless they are not too expensive and they can be assured that they are reliable. According to Coombs *et al.* (p. 8.15), tractor manufacturers find it difficult to recruit sufficient engineering and design staff to fulfill their present programmes. They also comment that 'the great need for reliability in agriculture is exercising a restraining influence on any rush to develop radical innovations which often tend to be unreliable in early stages' (p. 9.13).

Thus, the prospects for microelectronics-based automatic guidance systems for tractors and combines are likely to be very limited, at least for the next five to ten years.

SKILL IMPLICATIONS

Many of the problems experienced by British agricultural engineering firms are at least partly the result of their small size. Rothwell found that by contrast with British agricultural engineering firms, European firms nearly always have a Head of Department with a technical qualification, usually an engineering degree; the percentage of graduate engineers in research, design and development is generally higher in Europe; there is generally a higher proportion of technically qualified managers in European firms (Rothwell, 1979).

British firms also suffer from the disadvantage that they are not located in the midst of large areas of fairly homogeneous farming activity. The British market for agricultural engineering products is small and scattered; thus, for example, several small firms have manufactured electronic pig weighers, but they have had considerable difficulty setting up sales and service. Initial sales potential was low, so setting up their own sales and service network was uneconomic. Dealers were not interested in stocking products with small sales potential.

Problems of scale also afflict the small, specialist agricultural electronics firms. Perhaps the minimum viable team to develop microprocessor-based control systems for agricultural applications is about twelve: say, one systems analyst, one electronics engineer, one programmer plus back-up staff, mainly technicians. Even with a team of this size, important projects can be vulnerable to one person being sick or, worse, leaving the firm. Firms have had to drop projects because one person has left. Such firms, even if they attain such a critical size, have the problem of retaining their key electronics skills. As with other sectors of mechanical engineering, people with such skills generally prefer to work in the more technologically exciting areas of the electronics industry.

Rothwell found that many firms were suffering from skilled labour shortages, particularly toolmakers and fitters, which sometimes delayed production. This was often merely because the firms were located in rural areas, but was sometimes because larger firms nearby paid higher wages (p. 81).

REFERENCES

K.L. Blaxter, 'The Limits to Agricultural Improvement', Lecture, University of Newcastle Agricultural Society, December 1973, quoted in Rothwell, 1979.

R.W. Coombs, J.P. Gardiner, M. Gibbons, R.D. Johnston, *Incremental Innovation and the UK Tractor Industry*, Department of Liberal Studies in Science, Manchester University, 1978.

National Institute for Agricultural Engineering, Annual Report, 1977–78.

R. Rothwell, *Technical Change and Competitiveness in Agricultural Engineering: The Performance of the UK Industry*, SPRU Occasional Paper Series No. 9, September 1979.

8. Textile Machinery

P. Senker

Since the Second World War, innovation in the textile machinery industry has increasingly been the result of the incorporation of new, sophisticated techniques originating from outside the industry, including electronics. The main threat to the industry at present comes from Western European countries producing technically sophisticated, high quality machinery.

HISTORICAL BACKGROUND

The rapid transformation in the cotton industry, the first economically important example of machine industry, makes it the earliest, classical example of modern large-scale industry. Sporadic, localized and specialized application of machinery to textile manufacture occurred from 1598 onwards, when the stocking frame was invented by William Lee. Two inventions, Hargreave's spinning jenny and Arkwright's water frame were made at about the same time, in the 1760s, and were key elements in the industrial revolution of the following decades. By 1850, mechanization had spread to weaving and power looms predominated. Except for sewing machines, invented in 1851, in the textile industry proper, the period 1850–1950 produced few technical changes comparable to those of the earlier industrial revolution.

Between the two World Wars, the British textile industry declined dramatically and this was accompanied by a decline in the fortunes of the machinery industry. The period after the Second World War was one of radical innovation. The production rates of many types of textile machinery began to increase sharply after about 1950. The most radical innovation in spinning was open-end spinning, which can have a production rate of over three times that of conventional ring spinning. In this period, 'shuttleless' looms were introduced and labour saving, highly productive and sophisticated looms have been developed. Employment in the industry has declined from 75,000 in 1950 to 22,000 in 1979.

INDUSTRY STRUCTURE

Three firms dominate the industry, accounting for nearly half the industry's output. The top twenty-five firms account for about 80 per cent of the industry's net output. Between 1951 and 1973, the total number of establishments classified to the industry declined from 1,177 to 460.

In the early 1960s, the British weaving and knitting machinery industry declined dramatically. Between 1970 and 1975, one of the two major manufacturers of circular knitting machinery went out of business.

COMPETITIVENESS OF THE BRITISH INDUSTRY

The British textile machinery industry consistently exports about three-quarters of its output. Between 1954 and 1975, the British share of world trade in textile machinery fell from 30 per cent to about 11 per cent. After the Second World War much of the textile machinery manufactured in the United Kingdom was 'traditional', and exported to developing countries. This trend has been noticeable in other sectors of the engineering industry.

FACTORS AFFECTING COMPETITIVENESS

Low investment in technical change and poor quality have been as important as low capital investment and consequent low labour productivity as factors in the lack of British competitiveness. Until the mid-1960s, there was a lack of co-operation between British textile manufacturers and mill managers who were generally reluctant to test prototype machines. By contrast, Sulzer Bros. of Switzerland, who developed their radically new shuttleless loom in the 1930s, enjoyed the co-operation of a number of German mills.

Trading success has depended to a considerable degree on technical leadership. Several firms have declined from positions of dominance twenty-five years ago to near oblivion because they failed to produce new models. A study of over a hundred textile companies in the United Kingdom has shown that foreign-built machines have been acquired in preference to British because they are considered more reliable, efficient and productive (Rothwell, 1976).

The number of scientists and technologists employed by the industry fell by nearly 14 per cent between 1968 and 1973. During the same period, expenditure on research and development fell 30 per cent.

THE IMPACT OF MICROELECTRONICS

Computer control of jacquard knitting machines in particular and electronic mill monitoring have already had a substantial impact on the British textile machinery manufacturing industry. But electronics has been incorporated at every stage of the textile production process, from raw materials to finished cloth. For example, electronic controls have contributed substantially to the dramatic speeding up of the processes which produce yarn from synthetic material – extrusion, drawing out the yarn to provide elastic properties and texturing, from speeds of 20,000 rpm to nearly a million rpm today; and they have made it possible to combine these three processes into a single draw texturing machine which is cheaper to operate and which produces yarns of much higher quality and regularity.

McLean and Rush suggest that 'the use of electronics to achieve marginal increases in productivity at almost every stage of textile manufacture has been, whilst unspectacular in each individual case, extremely significant in its overall impact' (McLean and Rush, 1978, p. 24). McLean and Rush consider that the incorporation of microelectronics, and in particular microprocessors, is likely to proceed rapidly in the textile machinery industry, because this offers possibilities for further labour saving in textile manufacture and the past pattern of competition in the industry has been based on fast technological innovation.

DEC, an American firm without production facilities in the United Kingdom, have a virtual monopoly of the supply of computing equipment for both mill-monitoring and computerized knitting. (DEC are very important in the minicomputer industry: see Chapter 13.)

The few British manufacturers of textile machines which have survived the fierce technological competition of the past thirty years seem fairly well set up to withstand technological competition in the future. In particular, British manufacturers have not been slow to incorporate microelectronics into their products. But it is essential for British firms to have the capability to react quickly to radical innovations in textile machinery not directly based on electronic technology such as the Czech multiphase loom. There has been considerable growth in the number of consultants and systems manufacturers who provide an interface between the electronics and the textile machinery manufacturers. The computer controlled knitting machine – the major British innovation incorporating electronics into textile machinery – came about as a result of collaboration between electronics consulting firms and textile machinery manufacturers, although the particular firm concerned now has its own electronics R & D. It is possible that foreign manufacturers have been quicker to build up their own electronics R & D groups.

SKILL IMPLICATIONS

In his studies in the textile industry, Rothwell (1978) found several cases of failure of technical developments undertaken by consultants. For example, a firm lacking in-house fluidics experience used a local university as consultants. One reason for failure was that the company required to develop machinery which offered economic advantages to its customers: the consultants became engrossed in solving interesting technical problems, and although the machine developed worked well, it was much too expensive. A similar reason contributed to the failure of another firm's attempt to use lasers for automatic detection.

REFERENCES

M. McLean and H. Rush, *The Impact of Microelectronics on the UK: A Suggested Classification and Illustrative Case Studies*, SPRU Occasional Paper Series, No. 7, June 1978.

R. Rothwell, *Innovation in Textile Machinery: Some Significant Factors in Success and Failure*, SPRU Occasional Paper Series, No. 2, June 1976.

R. Rothwell, 'Some Problems of Technology Transfer into Industry: Examples from the Textile Machinery Sector', *IEEE Transactions on Engineering Management*, Vol. EM-25, No. 1, February 1978.

9. The Machine Tool Industry

P. Senker

Numerically controlled machine tools have been established products of the machine tool industry for many years. Computer controlled machine tools have been using microelectronic controls for several years. Cheap, reliable microelectronic components offer the possiblity of ever more sophisticated machine tool controls. But the British machine tool industry has concentrated on supplying the requirements of a relatively conservative home market, and on traditional Commonwealth markets. Research and Development effort has declined relative to that of other advanced industrial countries in recent years. Very few advanced DNC (direct numerical control) and no FMS (flexible manufacturing systems) have been supplied to the British market. The industry is very short of designers and the prospects of it applying microelectronics as successfully as major foreign competitors are very poor.

HISTORICAL BACKGROUND

The British technological lead in machine tools was lost some time in the latter half of the nineteenth century. By 1913, the British share of world trade in machine tools was about 12 per cent, compared with a German share of 48 per cent. By 1933, the German share had risen to 70 per cent, falling to 48 per cent in 1937. The British share gradually declined to 7 per cent by 1937 (Jones, 1078, pp. 12–13).

After the Second World War, Britain and America were the only countries with important machine tool industries which had not been devastated by the effects of war. In particular, the destruction and division of the German industry meant that the major world competitor in machine tools had been broken up, with the important machine tool region of Silesia in East Germany ('Machine Tool Report', 1978, p. 5).

In 1960, the Mitchell Committee reported that the performance of the United Kingdom industry suffered by its over-concentration on the production of standard machine tools. This accounted for the industry's shrinking share of world trade and for the high proportion of imports. This over-concentration on standard types was thought to spring from lack of sufficiently intensive development effort. For the same reason, the industry was over-dependent on licensing agreements to manufacture machine tools designed abroad and was also reluctant to meet the special needs of particular customers (Mitchell Committee, 1960, p. 27).

The machine tool industry is subject to extreme cyclical fluctuations. The severe recession in 1971 had a particularly adverse effect on the industry. Between 1968 and the last quarter of 1970, there was a strong boom. But by October 1970, manufacturers were concerned that they were facing a recession (Bolter, 1970a). By February 1971, it was clear from the figures for home market orders for the last quarter of 1970 that a recession was well under way (Dwyer, 1971a; Keegan, 1971). The recession intensified during 1971 (Dwyer, 1971b; Rodgers, 1971).

The results of the recession, combined with the recent collapse of Rolls-Royce, were very severe in terms of redundancies. On 5 February, Alfred Herbert announced 1,000 redundancies (Corina, 1971). In the succeeding twelve months, Herbert's cut their labour force from 11,000 to 7,000, reduced their number of plants from seventeen to eleven and reduced their product range from 300 to 200. In the twelve months to November 1971, 18,000 workers left the machine tool industry (Dwyer, 1972).

Staveley Industries were already in trouble before the recession really bit. In the mid-1960s, Staveley, according to the chairman appointed in 1970, had made massive 'over-ambitious' acquisitions of machine tool companies 'without sufficient management and manpower resources to ensure the smooth integration of the newly acquired machine tool companies and their products' (Dwyer, 1970). Towards the end of 1970, Staveley made 700 men redundant at their Craven-Swift factory (Bolter, 1970b). Between early 1970 and 1975, Staveley reduced its total machine tool employment from 5,000 to 2,500 (Gooding, 1975).

In the year ending November 1972, Newall Machine Tool reduced its total employment from 1,750 to 1,350 (Gooding, 1973).

Tube Investments entered the machine tool business with the money received when its steel interests were nationalized. First, Charles Churchill and Co. Ltd. was acquired and then Vero Machine Tool Co. Ltd. and the Froriep GmbH factories in Germany. In 1969, Tube Investments, with a machine tool workforce of 3,000, acquired Coventry Gauge and Tool Co. Ltd., which then employed 2,500 (Topping, 1969). In April 1971, 259 redundancies at Coventry Gauge's Coventry plant were announced (Cassell, 1971).

Employment in the industry fluctuated between about 80,000 and 90,000 in the 1960s, but has declined steeply in the 1970s to the present level of under 65,000.

INDUSTRY STRUCTURE

The Way Committee reported in late 1970 that, since the Mitchell Committee Report in 1960, the number of machine tool enterprises had fallen from 350 to 200. In 1960, twenty machine tool companies supplied half the industry's deliveries. By 1970, seven machine tool companies accounted for a similar proportion of deliveries. An improvement in performance in 1969 suggested to the Way Committee that restructuring may have been beneficial (Way Committee, 1970, pp. 16–17). The seven major groups in the industry are shown in Table 9.1.

It is quite possible that concentration of the industry was a necessary condition for future specialization and for the industry's future prosperity. But it is clear the concentration by itself did not improve the industry's performance.

TABLE 9.1 Structure of the machine tool industry

Group	*Some constituent companies*	*1976 turnover £m.*
John Brown	Wickman	48
Cincinnati Milacron		32
B. Elliott	Elliott, Butler, Snow, Newall (acquired 1977)	57
Alfred Herbert		49
Staveley	Asquith, Drummond, Hearns-Richard, Staveley-Laporte	27
600 Group	Colchester Lathe, W.E. Skykes, F.J. Edwards	51
Tube Investments	Matrix, TI Churchill, Coventry Gauge	42

Source: *Financial Times*, 20 September 1977, quoted in 'Machine Tool Report', 1978, p. 7.

COMPETITIVENESS OF THE BRITISH INDUSTRY

Since the early 1970s output in the machine tool industry has stagnated or declined in most countries and employment has fallen substantially. West Germany has been the dominant exporter of machine tools throughout this century, currently accounting for about 40 per cent of OECD exports. The United States and Switzerland are next with 12–13 per cent (Jones, 1979). Between the mid-1950s and 1976, the British share of world machine tool markets fell by nearly a half – from 10½ per cent to 5½ per cent. In the 1950s and 1960s, British world ranking as a machine tool producer varied between 3rd and 5th. By 1976, it had fallen to 7th (Coventry, 1978, Table 3).

The British share of world production of machine tools has shown a similar pattern. Between 1960 and 1976, the British share of world machine tool production fell from over 13 per cent to 8½ per cent. The American share also fell – from 39 per cent to 27 per cent, but West Germany (25 per cent to 32 per cent), Japan (8 per cent to 14 per cent) and Italy (6 per cent to 9½ per cent) all increased their shares, with the French share first falling and then regaining their 1960 share of about 8 per cent (Coventry, 1978, Table 4).

The British share of markets for the more advanced machine tools has fallen most sharply. This is illustrated by the decline of the proportion of British exports going to advanced industrial countries. Between 1971 and 1976, the proportion of British machine tool exports going to advanced industrial countries declined from just under 85 per cent to just over 70 per cent (Coventry, 1978, Table 9).

The degree to which Britain has fallen behind technologically is also illustrated by the much lower average unit price of machine tools produced in Britain than in West Germany. West German machine tools are half as expensive again per unit as British machine tools. Amongst the more advanced specialized type of machine tools, the differences are even more marked. American advanced machine tools are as expensive as West German (Jones, 1978, p. 19).

British imports of numerically controlled machine tools have risen greatly. Between 1971 and 1976, British imports rose from 34 per cent to 50 per cent as a percentage of home consumption. During this period, numerically controlled machine tools rose from 6 per cent of all British machine tool imports to 17½ per cent – a massive rise from £2.7 million to £23.5 million (Coventry, 1978, Table 1 and pp. 2–3).

But Britain also faces serious competitive threats in the markets for standardized machine tools, particularly from the newly industrializing countries. In addition, American standardized machine tools are slightly cheaper than British (Jones, 1978, p. 19).

FACTORS AFFECTING COMPETITIVENESS

Pavitt and Soete have shown that there is a particularly strong statistical relationship between the innovativeness of OECD countries' machine tool industries – measured in terms of their patenting activity – and their per capita exports. Thus, there is a reasonable presumption that innovation is a significant factor in the international competitiveness of a country's machine tool industry (Pavitt, 1980).

Between 1955 and 1967, expenditure on R & D increased very fast (see Table 9.2). In the period 1958–62, the British share of world machine tool trade fell to around 6–8 per cent. But, by 1968, the British share had recovered to 10 per cent – and it is possible that the sharp increase in R & D played some part in making this possible. Indeed, in 1970, the Way Committee Report suggested that the Mitchell Committee's criticism

TABLE 9.2 Research and Development activity in the metalworking machine tool industry, 1955–1975*

	1955	1967	1972	1975	Change 1967–1975 (%)	
	R & D employment ('000)				R & D empl.	Total empl.
Machine tools	0.6	1.8	0.4	4.3	–61	–24
Mechanical engineering	12	19	9	10	–48	–11
Manufacturing	110	196	162	159	–18	–10
	R & D expenditure (£m 1975 prices)†				R & D expend†	Total output‡
Machine tools	3.3	14.1	2.7	4.3	–69	–12
Mechanical engineering	62	135	63	63	–53	+18
Manufacturing	427	1421	1264	1165	–18	+14

*Expenditure carried out within and employment in private industry only.
†Expenditure at current prices adjusted to 1975 prices using a combined index of capital goods prices and hourly earnings in manufacturing.
‡Industrial production.
Source: Jones, 1978, p.21.

in 1960 of the machine tool industry's neglect of applied research and development no longer applied with the same force, in view of the increase in applied R & D effort between 1964–5 and 1968–9. (Way Committee, 1970, pp. 66–70). However, the subsequent precipitous decline in R & D effort may well have been an important factor in the subsequent decline in the industry's international competitiveness.

THE IMPACT OF MICROELECTRONICS

The principal impact of microelectronics on machine tool manufacturers is to expand the area of application of numerical control. At the same time, the introduction of microprocessors into NC has made possible cheaper and more reliable systems, which usually offer more features at the same price as the equivalent hard-wired systems.

NC machines first became available in the early 1950s. By the mid-1960s, a large range of NC systems had been developed. By 1970, numerical control systems could provide all the basic control systems for both traditional machine tools and for more complex machining centres which had been developed specially to allow the maximum number of different machining operations on a component at a single setting. At this time, systems were all hard-wired and special purpose. Separate controllers had to be designed for each type of machine.

The hard-wiring and lack of flexibility meant that input data had to be specially coded by a programmer. Larger jobs required computer-aided part programming. The need for large computers for part programming tended to concentrate the use of NC in the larger, more sophisticated companies, although programming was available through computer bureaux. By about 1970, minicomputer prices had fallen to the point where it was economic to build an NC system around a standard minicomputer.

1. Computer Numerical Control (CNC)

The main feature of a CNC system is its flexibility. The features provided can be changed by altering the control programme. Much more extensive diagnostics – self-checking tests during normal running, and special programmes for maintenance and test purposes – can be provided by CNC systems.

All new NC systems now incorporate microprocessors. In the last two or three years, two types of microprocessor CNC have developed:

(a) Fully flexible: these are virtually the same as minicomputer CNC systems but use LSI chips;

(b) Fixed-feature: these are usually offered with standard specifications designed to meet the needs of most users at the lowest possible cost. These are modern equivalents of hard-wired systems, but offer additional features such as part programme storage and editing, and backlash compensation.

2. Direct Numerical Control (DNC)

Direct numerical control involves linking groups of machine tools to a central computer. Information, e.g. work completed, interruptions in production, is fed back to the computer and can be used for scheduling and production control by the computer.

In 1978, there were about a hundred DNC systems in the United States and perhaps about eighty in Japan. It was suggested that there were three in Britain (Hollingum, 1978). For a time, DNC was considered by many to be uneconomic. They felt that for most practical purposes CNC, minicomputer or microprocessor control of individual machines, was more useful. But it has been suggested recently that DNC is now being reconsidered in conjunction with computer aided design. Recent systems operate by the central computer distributing complete NC programmes to CNC controllers at the individual machines.

3. Flexible Manufacturing Systems (FMS)

A flexible manufacturing system is a group of DNC machines controlled by a computer in which workpieces are transported between machines automatically. There are about twenty FMS systems operating in Japan, a few in the United States and none in Britain.

The nearest all-British products to the intermediate systems now being produced in the United States and Japan are machining centres being produced by Kearney & Trecker Marwin (Hollingum, 1978). In 1978, machine tool companies were being asked to tender for a one-year design study for a new generation of automated small-batch production machines. Hollingum's review of the ASP project (1978) suggests strongly that the industry's reluctance to tender for the study is a result of its grave shortage of design engineers insufficient even to meet their current short-term commitments.

4. Sheet Metalworking

The machine tools considered so far are metalcutting machines. But CNC control is also being applied to sheet metalworking. Overseas manufacturers dominate markets for such machines.

SKILL IMPLICATIONS

The machine tool industry has been described as 'one that predominantly relies on engineering skills'. There is little doubt that the British machine tool industry's long-term lack of competitiveness is to be traced to its inability to develop and utilize the skills and scientific knowledge necessary to remain in the forefront of technology on a world-wide scale.

> The delay in adopting the most advanced designs and techniques is attributed to the failure of the industry to attract sufficient skilled engineers, its reluctance to devote sufficient resources to research and development and the lack of attention paid to the problems of production in the design of new machine tools, insufficient emphasis placed on marketing and after-sales service, the specific requirements of customers and so forth (Jones, 1978. p. 19).

Jones observes 'A continuation of the trends observed in the 1970s, without any major change in the amount of government intervention and at the same time maintaining open markets will almost certainly result in a further run down of the industry' (Jones, 1979).

The Mitchell Committee had surveyed the top firms in the industry, employing about 30,000 people (about two-thirds of industry employment) and found that only twenty-five graduate engineers were employed. In the late 1950s only 120 qualified scientists and engineers were employed in Research and Development in the industry.

Traditionally, machine tool designers are recruited from the ranks of draughtsmen. In a recent study (Swords-Isherwood and Senker, 1978), a machine tool firm outlined the normal progression as draughtsman to machine tool designer to senior machine tool designer to section leader. Mechanical engineers are sent on postgraduate M.Sc. courses, but the payoff from this type of training programme is considered to be very long term.

Both machine tool firms interviewed in our study of shops (Swords-Isherwood and Senker, 1978) had problems in relation to shopfloor and production engineering skills. One firm complained of a lack of shopfloor skills which they attributed largely to inadequate recruitment of apprentices in the previous several years. This, combined with a high average age of the workforce, had induced them into a major skill training programme. Like many firms in the industry, this firm had been involved in mergers, redundancies and reorganizations. At various stages, these had caused a loss of skilled men whom they now need. For example, during one reorganization, the firm missed recruiting as many apprentices as they would have wished by waiting for permission to recruit from the firm's new owners. This, by itself, may be trivial. But it is not surprising that an industry which, in a series of short term measures, periodically cuts back recruitment of apprentices, makes its skilled men redundant and undergoes frequent reorganization, finds that it is short of skilled labour.

Another firm's dilemma expressed clearly traditional problems of the British engineering industry in relation to highly qualified manpower. In order to get the best out of their own numerically controlled machine tools, they felt a need for first class practitioners in every area – e.g. for turning, milling, boring. But they thought that it was essential for such men to be time-served craftsmen in their particular area of expertise, as well as knowledgeable of other related processes. They felt that a graduate of a British university would be incapable of doing the job, unless he had also served an apprenticeship in a major machine tool company. It was felt that a similar situation would apply in the United States, France, Italy and Spain, but that in West Germany, where technical education is 'closer to the shopfloor', a graduate would probably be capable of doing the job. Further, the salary which they were able to offer a 'machine tool project engineer' would not be very great, as the works manager's salary was itself relatively low. The result of the inability to attract appropriately trained people was failure to get the best out of the numerically controlled machine tools which had been installed.

This analysis points to the short-term basis of the British industry's decision making in relation to manpower and skills. In the late 1960s, it is possible that most of the industry's managerial effort was directed to restructuring, rather than to intensifying technical effort. In the recession beginning late in 1970, the industry appears to have cut back both on its research and development staff, which had been building up until the mid-1960s, and on its apprentice recruitment. Not surprisingly when demand began to pick up again, the industry continued to be affected by skilled labour shortages. Probably more important, the failure to sustain its research and development effort has led to a continuation of the long-term decline in its international competitiveness. The manufacture and marketing of flexible manufacturing systems embodying microelectronics may not be critical to the inudstry's future survival and prosperity quite so quickly as microelectronics is affecting the products of other sections of the mechanical engineering industry (e.g. materials handling, textile machinery).

REFERENCES

H. Bolter, 'Machine tool orders up but makers concerned', *Financial Times*, 8 October 1970a.

H. Bolter, '700 men to lose their jobs', *Financial Times*, 27 August 1970b.

M. Cassell, 'Machine tools: now TI to dismiss 250' *Financial Times*, 30 April 1971.

M. Corina, 'Redundancy rises as industry makes massive cutbacks', *The Times*, 15 March 1971.

Coventry Machine Tools Workers' Committee, 'The Fight for Jobs in the Machine Tool Industry', 1978, a report commissioned by the Committee.

D. Dwyer, 'Machine tools: largest group find it tough going', *The Times*, 4 March 1970.

D. Dwyer, '20.7m drop in new orders for machine tools', *The Times*, 18 February 1971a.

D. Dwyer, 'Another big fall likely in machine tool orders', *The Times*, 21 June 1971b.

D. Dwyer, 'Machine tools have still to beat the cyclical curse', *The Times*, 29 March 1972.

K. Gooding, 'Product development programme gets under way as profits recover', *Financial Times*, 17 September 1973.

K. Gooding, 'Fighting on a reduced front', *Financial Times*, 22 April 1975.

J. Hollingum, 'Machines to change the face of batch production', *The Engineer*, 13 July 1978.

D. Jones, 'Plant Size and Efficiency in the Production of Metalworking Machine Tools', NIESR Discussion Paper, July 1978, pp. 12–13, cites, among other references the Mitchell Committee Report, 1960, p. 23.

D. Jones, 'Intervention and Adjustment in the Metal Working Machine Tool Industry in Western Europe – A Draft Outline', mimeo, 30 January 1979.

V. Keegan, 'Machine tools trade worst for 4 years', *The Guardian*, 18 February 1971.

'Machine Tool Report': a report commissioned by the Joint Shop Stewards Committee of TI Churchill Ltd., Blaydon-on-Tyne, from the North East Trade Union Studies Information Unit, 1978.

Mitchell Committee Report, Board of Trade, *The machine tool industry*, HMSO, London, 1960.

K. Pavitt, ed., *Technical Innovation and British Economic Performance*, to be published by Macmillan, 1980.

P. Rodgers, 'Slump in machine tool orders gets more severe', *The Guardian*, 9 September 1971.

N. Swords-Isherwood and P. Senker, 'Technological and Organisational Change in Machine Shops', Science Policy Research Unit, University of Sussex, mimeo, May 1978.

D. Topping, 'TI strengthens machine tool base in £6.4m Coventry bid', *The Times*, 13 May 1969.

Way Committee Report, Department of Trade and Industry, *The machine tool industry*, HMSO, London, 1970.

10. Welding Equipment

C. Huggett

The manufacture of welding equipment is a small, but important, part of the engineering industry. It is of particular interest here as microelectronics has already been applied extensively in resistance welding for mass production and is likely to be applied extensively to arc welding. This chapter is concerned mainly with these two areas of application.

HISTORICAL BACKGROUND

Until about 1900, welding techniques were relatively crude and cumbersome, but at that time the three basic techniques of modern welding were developed: arc welding, resistance welding and gas welding. The First World War, with its demand for transport and other shipping, led to the development of arc-welded ship construction. From around the same time, expanding oil-field operations, coupled with the growth of oil-refining and other large-scale chemical processing, reinforced the growing use of arc welding. The expansion of production of consumer durable goods made from steel sheet led to a growth in the use of resistance welding. Thus the use of arc welding has been heavily concentrated in the 'heavy' engineering sectors, and resistance welding in the high-volume automobile and appliance sectors.

As different joining needs have arisen, new variations of the basic processes, together with a range of novel techniques, have been developed. In recent years technological effort has focused on developing quite new methods (e.g. electron-beam, friction and laser welding), but the main emphasis has probably been directed towards increasing the quality, reliability and precision of welding and at increasing welding speeds. At the same time, substantial efforts have been made over the last decade to substitute equipment ('robots') for the manual control and manipulation of welding devices (see Chapter 4).

STRUCTURE OF THE INDUSTRY

The supply of welding equipment involves production at three levels:

Level I. System engineering and integration.

The design of large-scale, relatively complex systems, which involve welding equipment as only one major element.

Level II. Production of basic system elements.

Welding systems may involve the use of one or more of the following four basic elements:

(a) The welding devices themselves (e.g. welding guns, multi-spot welders etc. plus the associated power packs and immediate control systems);

(b) 'Robots' (devices to hold the welding gun and move it through the required sequence of functions);

(c) Electronic control systems (programmable devices to control other elements in the system, and perhaps to monitor and record various aspects of operating performance);

(d) Transfer and handling devices. (These include all the conveyors, jigs, workpiece-locating devices etc., required to move workpieces through the various welding stations and at each station to hold workpieces in precise locations. Robots for workpiece handling may be incorporated.)

Level III. The supply of parts and components for the main elements produced at Level II.

Production of the different product groups of the industry is organized in slightly different ways, and the organization of the industry in the United Kingdom appears to be different from that in other European countries and in Japan. Production and supply of high-volume resistance welding equipment requires a substantial degree of integration between activities at Levels I and II, and between the different types of production within Level II. Arc-welding systems tend to be less complex, with a lower requirement for integrating the different levels and types of production – although integration at Level II between the production of robots and of welding devices is becoming increasingly important.

In a variety of ways the organization of the industry in continental European countries reflects these requirements for interaction and integration – particularly with respect to more complex, high-volume, resistance welding equipment. For example, Sciaky (France) has designed and engineered integrated, high-volume resistance welding lines for many European automobile companies. It supplies its own welding equipment for these lines and, with experience in the production of more conventional multi-welder transfer lines, also produces its own transfer system elements. The company often follows a policy of designing its own special-purpose robots for particular applications, rather than incorporating standard robots from specialist suppliers.

KUKA (Germany), a subsidiary of IWKA, is another specialist supplier of integrated resistance welding lines, but draws more heavily than Sciaky on other companies for the basic elements involved. More recently, the supply of basic elements has become more integrated with its own systems engineering, and it has begun producing its own robots, which incorporate control systems from Roth Electric (another IWKA subsidiary).

Both KUKA and Sciaky entered the market for high-volume resistance welding lines because of their earlier activities as suppliers of welding equipment. Unlike these companies, ASEA, a Swedish supplier, has entered the market from a base in robot system manufacturing. Jointly with ESAB, a Swedish supplier of welding equipment, it has engineered robot spot-welding lines for Saab-Scania.

Japanese automobile companies seem to rely relatively heavily on their own systems engineering. But, in addition to drawing on specialized producers of robots (e.g. from Nachi,

Toshiba and Kawasaki), they also incorporate robots made within their own, often highly diversified, groups, e.g. Mitsubishi.

Another significant feature of the organization of the industry in continental Europe and Japan is the role of the equipment user. In a number of cases users have themsleves carried out the overall design and engineering of systems for their own use, e.g. Volvo and Daimler-Benz. In other cases the role of the user goes more deeply into the supply structure at Level II. For example, Fiat has relied heavily on its own subsidiary company (Comau) for the supply of welding lines. In some instances (e.g. Trallfa), entry by companies into the market has been based on the initial development of major system elements (robots in this case) for their own use. In Japan also, welding equipment users have played a significant role in developing more advanced types of system; although here they tend to be large organizations with wide-ranging interests (e.g. Kawasaki, Mitsubishi and Kobe Steel).

The structure of the British industry does not exhibit these features to the extent evident in other countries. There is a small number of relatively specialized major producers of welding equipment, plus a number of smaller firms in the industry. None of these seems to have very close structural links either with equipment users or with suppliers of other major elements that need to be integrated into more advanced welding systems. However, Hall Automation Ltd. (HAL), until 1979 the only British producer of robots for welding applications, has operated with close working links to the British Oxygen Company (BOC) – a welding equipment supplier. HAL has recently announced its merger with the large General Electric Company (GEC).

COMPETITIVENESS OF THE BRITISH INDUSTRY

Changes and limitations in the nature of available statistics prevent a clear and detailed analysis of the competitive position of the British industry. However, for the period 1975–77, it is possible to *estimate* that the United Kingdom had a positive net balance of trade in welding equipment: the value of exports was almost one and a half times as large as the value of imports (Huggett and Bell, 1979). Moreover, this situation seems to have improved over the period: the rate of growth of the value of exports during the three years (115 per cent) exceeded that of imports (98 per cent), and substantially exceeded that of total production by the sector. In consequence, the export intensity of British production seems to have been rising quite rapidly during the period – roughly doubling from about 20–30 per cent to about 35–40 per cent.

However, total sales have grown much less rapidly than imports. Indeed, net of the rapidly growing exports, total British production may have fallen slightly, while imports have almost doubled. This suggests that, despite the export growth during this period, the British industry was not meeting the needs of British users of welding equipment as effectively as foreign competitors.

Perhaps more than one third of the value of British investment in welding machinery over the three years consisted of imported equipment. On its own this may not seem to be a matter of great concern, but it raises a second point. The recent level of import intensity in the value of new investment in welding equipment can be compared with the level of import intensity in the *numerical stock* of welding machines already in use in the engineering industry. The Metalworking Production Census of Machine Tools for 1976 shows that about 93 per cent of the number of installed welding machines had been supplied by British producers. There is, therefore, a very big discrepancy between the proportion of the *numbers* of welding machines already installed in 1976 which consisted of imports (about 7 per cent) and the imported proportion of the *value* of additions to that stock which were being made in the years immediately around 1976 (about 35 per cent). This discrepancy may be the result of some combination of (a) a recent and very rapid rise in the volume of imported welding equipment and (b) a tendency for imports to be very heavily concentrated in more advanced, higher unit-value types of welding equipment.

Declining competitiveness, suggested by analysis of the import situation, is not inconsistent with the apparently healthy position in relation to exports. Exports may consist of less advanced and complex types of equipment, and their rapid growth during 1975–77 may even, in part, be a response to a weakening position in the domestic market.

An impression of the recent competitive position of the

British industry in the market for more advanced welding systems can be derived from a review of two specific markets in that category: the market for high-volume, resistance welding equipment and the market for robot-controlled arc-welding equipment.

(a) High-volume, Resistance Welding

Within the British industry there is no company like Sciaky which has supplied complete welding systems to an international market. No British automobile producer has, like Fiat, developed welding systems (or major elements of them) for supply to other users. While companies like Volvo, Daimler and Mitsubishi have played significant roles in the development of such systems, no British automobile producer seems to have played a similar role. The competitors for supplying BL with new high-volume welding lines in the late 1970s apparently did not include a British company. Thus, in the United Kingdom, neither specialist producers of welding equipment or of other basic elements for welding systems, nor system suppliers based on expertise as users, seem to be competing at Level I – the development, design and supply of integrated systems.

British companies specializing in the production of basic system elements (Level II) might compete as suppliers of these to other companies producing the integrated systems. For example, Ford has used Comau system elements, while Fiat's own lines supplied to Volvo used Unimate robots, and KUKA used Japanese (Nachi) robots for some lines. No British company seems to have emerged as such a specialized supplier – either to international markets or even to foreign companies building welding lines for British plants.

The situation may be different at Level III: the supply of parts and components to the foreign producers of welding systems or their basic elements. KUKA, for example, in supplying welding lines to British companies (e.g. BL and Ford UK) has drawn on British suppliers for a proportion of the parts and components such as motors, hydraulic systems, switches and controls. KUKA has also been reported as seeking British suppliers of parts and components for lines other than those they install in the United Kingdom, apparently because British costs for such items are lower than in Germany. This competitive capability to supply parts and components owes little to technological leadership. Rather it reflects the type of capability which attracts component assemblers to developing countries in Asia and component manufacturers to subcontractors in North Africa or India.

Thus, in the market for high-volume, resistance welding equipment which incorporates advanced electronics technology, this review would suggest that the most promising role for the British engineering industry may be as a high-class jobbing shop for other European capital goods producers.

(b) Robot-controlled Arc-welding Equipment

By the early 1970s, British industry was active, if somewhat belatedly, in the first phase of development of products in this market. This phase involved the linking together of standard arc-welding equipment and standard general-purpose robots. Both GKN and Hawker Siddeley Dynamics (HSD) were involved. However, as in other countries, not very much that was immediately useful seems to have emerged from this phase of development. The main output was probably accumulated experience, which pointed towards the need for more *special-purpose* development of 'automated' arc-welding systems. Neither GKN nor HSD moved into this second phase which gathered momentum from around 1973–74. Fortunately, the knowledge of the robotics team at HSD was not totally dissipated, since the personnel moved to join a new, small company (Hall Automation) which was set up in 1974 by the former leader of the HSD team. This company developed links with the British Oxygen Company which had experience in the arc-welding area. For the rest of the 1970s the competitive position of the United Kingdom in this market rested on the progress of this single small firm. Having abandoned to foreign competitors the development of robots for welding (and other) applications, the large corporate sector of British industry stood aside from direct commitment of resources to product development for this market. Only some six years later, with the risks and uncertainties considerably reduced, was the British competitive strength in this market substantially backed by the resources of a large industrial firm.

By the time of the international welding exhibition at Essen in 1977, a prototype product was on show but 'BOC said that it has yet to carry out more development work on the mechanics of the manipulator and weld positioner' (Page, 1977). The microprocessor-based control system was much admired. The company claimed that in some respects, it 'had few, if any rivals', and the system was 'awaited with keen interest'.

By August 1978, Hall Automation was completing check-out of its control systems and was reported to be ready 'to go ahead with delivery of the first five arc-welding robots to BOC', (*The Engineer*, 1978), while BOC was '*planning to* set up a demonstration area at its . . . factory to show the machine on customers' own work'.

By mid-1979, six of the BOC-HAL systems had been sold, and output was scheduled to run at about one per month. Output of *all* types of HAL robot systems was, at that stage, planned to rise to about three per week by late 1981 (Kerr, 1979), but those destined for arc-welding applications only constituted about a third of total output from the small company – with employment still less than thirty in 1979.

Thus, British industry is active in this market, but only just – with production planned to reach a rate of about fifty units per year in 1981 – some ten years after British work in this area started. However, already by 1976, a range of arc-welding robot systems had appeared from European and other companies (Unimate, KUKA-Nachi, ESAB/Trallfa, ESAB-ASEA, ARO, etc.). ESAB had installed its first in 1974, and by 1976 claimed to have about thirty in action in Europe, while by the same date KUKA claimed to have about twenty-six of its Nachi-based systems in action in Japan and Europe.

By 1977, ESAB claimed to have twenty-five of its Trallfa-based arc-welding robots in use in Norway alone. By mid-1978, ASEA had about thirty systems in use in Europe, and Unimation had a similar number in use throughout the United States and Europe. Competition was increasing as future markets were expanding. At least one major new entrant had appeared – Cincinatti-Milacron with its microprocessor-controlled 'Acramatic' system for MIG/CO_2 welding.

By mid-1979, Shin-Meiwa (Japan), having started work on developing and designing a special purpose robot system for arc welding in 1973 (after earlier trials with general-purpose robots), was manufacturing a range of four rather simple welding robots. It was reported as having produced more than a hundred of these, and Kawasaki had sold twenty point-to-point systems for arc welding. Trallfa (Norway) had sold over 600 robots, and about 5 per cent of the output of this small company was scheduled for arc-welding applications. Kobe Steel had entered the market with its microprocessor-controlled 'Arcman', and another American welding equipment company (AiResearch) was moving into the robot arc-welding field with its Unimate-based system.

The international market for robot-controlled arc-welding equipment is rapidly expanding. The greater accumulated experience of production, marketing and equipment-use that is available by now to other companies may limit the ability of this British firm to exploit this market. However, the recently announced merger of HAL with GEC may provide the resources needed to consolidate the competitive 'toe-hold' of British industry.

FACTORS AFFECTING COMPETITIVENESS

Among the many factors which contribute to competitive success in the international market for welding equipment, one deserves particular emphasis: the involvement of users in the development of advanced forms of welding equipment.

The commitment of an adequate scale of resources to product development will be required for competitive performance in this sector. Perhaps even more important will be the persistent determination to push through the risks and uncertainties of this phase of competition. There may be grounds for thinking that the British welding equipment industry has failed to allocate to product development over the past ten years adequate resources to ensure a strong competitive position in the welding equipment market. Others seem to have pushed the product technology forward more effectively. However, it also seems very apparent that British *users* of welding equipment have not been pulling very hard on the equipment suppliers.

We have already noted the significant role in product development which has been played by users of welding equipment in other economies. One might also note the role of users in the development of related capital goods. Unimation has developed a new, low-cost robot for precision assembly: the PUMA. This development was carried out over several years in close collaboration with General Motors, and that company is reported already to be using about two-thirds of the thirty or so PUMAs which have been sold (Marsh, 1980).

A similar pattern is evident in the embryonic market for image-sensing assembly robots. Both IBM and Texas Instruments are reported to be entering the market on the basis of devices which they had developed initially for use in their own production operations (Marsh, 1980).

Close interaction between system-use and system-production provides the innovative capital goods firm with three types of competitive advantage:

(1) it provides the firm's finance directors and bankers with tangible evidence to support the persistent allocation of resources at an adequate scale;

(2) it provides the firm's engineers with the 'learning' experience in the *use* of new systems which is essential for successful product development;

(3) it provides the firm's salesmen with practical, working, demonstrable products.

However, the apparent failure of the British industry to capture these kinds of competitive advantage cannot simply be attributed to the firms involved. There is little prospect of significant user-involvement in the development of innovative capital goods like welding equipment if such users operate in an economic environment which does not generate pressures for the introduction of such equipment.

In a high wage and high productivity economy, the pressure to seek innovative capital goods will be significant. In a high growth economy, the rate of investment will feed back to provide economies of scale at an early stage in the production of new capital goods. One or both of these conditions were part of the context of the French, German, Scandinavian and Japanese producers of welding equipment during the 1970s. Neither was a feature of the environment of British producers. The significance of the former may be reflected in the limited involvement of equipment users in the development of new welding systems. The importance of the latter may be reflected in some of the problems faced by Hall Automation: for example in 1979 it was having to buy individually-pressed mild steel covers costing around £500 apiece instead of plastic mouldings which if supplied in volume would cost about £100 each (Kerr, 1979).

For the British firms in this sector of the capital goods industry, domestic demand for the development of new types of welding equipment lagged behind the demand arising in other economies. For their competitors the necessary interaction between system-user and system-producer would arise 'naturally' before it was a practical possibility in the United Kingdom. This situation provides competitive advantages in international markets to non-British firms which can more easily and rapidly capture the 'learning' and scale economies involved in introducing innovative products.

THE IMPACT OF MICROELECTRONICS

The growing use of electronic systems within welding equipment has been outlined in Chapter 4. Here we need only note three particular points.

(1) Electronic systems constitute only a part of any unit of welding equipment. Even for the most basic welding device, electronic elements will constitute only part of the immediate control system. For more complex welding systems (e.g. robot-controlled arc-welding equipment or high-volume resistance welding lines) a very large proportion of what is involved remains based on mechanical, hydraulic, and electrical engineering.

(2) The advent of microelectronics in particular does not seem to have opened up any fundamentally new product concepts which were not already practical realities or well-developed plans in the pre-micro era of electronics development. By making such concepts cheaper to realize and more convenient and reliable to use, microelectronics has probably speeded up and diversified the development and application of electronic control systems for welding equipment.

(3) Further significant developments towards the automation of welding – especially of arc welding – depend not so much on fuller or better utilization of microelectronics, but on the development of devices to track and monitor the progress of the welding process itself at the interface between the components to be joined and the immediate source of the power (and filler) and which joins them.

SKILL IMPLICATIONS

Clearly the skills required to develop and produce electronic systems will be required for competitive performance in the welding equipment industry. Competence in this area will be a necessary condition for business success. However, this should not induce any kind of microelectronics myopia. As noted above, even the most electronics-intensive welding system is still a *welding* system based on the design and production skills involved in electrical, mechanical and hydraulic engineering, as well as those involved in electronics. The development, production and marketing of such systems requires engineering and business skills across the board. One should note that the apparently growing import-intensity of British investment in welding equipment was probably underway before advanced electronics, let alone microelectronics, was a very significant feature of product competition.

REFERENCES

The Engineer, 'Robot maker set for solid state "micros" ', 17 August 1978, p. 15.

C.M. Huggett and R.M. Bell, 'Microelectronics, welding and skills: a review of some trends in the engineering industry', SPRU, mimeo, 1979.

J. Kerr, 'A promising outlook for Britain's only robot manufacturer', *The Engineer*, 2/9 August 1979.

P. Marsh, 'Electronics giants see the light with robots', *New Scientist*, 28 February 1980, p. 641.

M. Page, 'The light-weight welder with lots of power and bite', *The Engineer*, 13 October 1977, p. 21.

PART II INDUSTRIAL SECTORS

B. The Electronics and Allied Industries

Skills are becoming increasingly similar in all sectors of the electronic equipment industries. Two factors account for this: first, the increasing share of systems (both in terms of numbers of functions and in terms of value-added) taken up by microelectronics; and, second, the technical convergence within the information technologies – data processing, word processing and communications. For these reasons, although analysis of technical change and competitiveness in three sectors of the industry (Telephone Equipment, Televisions and Minicomputers) are presented separately below as in Part IIA, the implications for skill are considered together for these industries in a separate chapter. This skill implications chapter also draws on several interviews by ourselves and by BPA Marketing and Management in companies throughout the industry.

11. Telephone Equipment

P. Senker

Between 1971 and 1979, employment in the manufacture of telephone and telegraph equipment fell sharply from 88,700 to 67,600. Changes in the technology of telephone equipment, falling shares of export markets and policy changes, especially in relation to the British Post Office (BPO) combined to cause this reduction. In 1963, the British industry held 25 per cent of world telecommunications export markets. This has now fallen to 6 per cent (Manasian, 1979). Inevitably, this has been reflected in widespread redundancy.

HISTORICAL BACKGROUND

Manual Systems

Manual systems are by now almost entirely obsolete. The basic form of switch was a board with an array of sockets, each enabling a connection to be made to a different line. The operator could connect any two lines together by means of a cord, plugging one end into each socket; many such connections could exist simultaneously on a single board. The manual board was a simple but powerful switching device. Up to 10,000 lines were connected to one board worked by a team of operators.

The major constraint on obtaining rapid and reliable connections with manual techniques was the number of operators required. Manual systems had some advantages over later automatic techniques. Signals between operators and subscribers were mainly in the form of speech. Human operators could carry out requests – such as trying to make connections later which could not be made at the first attempts – which were difficult, or impossible, for the automatic machinery that superseded them.

Automatic Systems

(1) Strowger

Strowger equipment was first installed in America in 1892. It remains in widespread use especially in the United Kingdom. Its original purpose was to eliminate the need for human operators in setting up calls.

Strowger equipment, working on electromechanical principles, selects appropriate connections, checks whether lines are free and makes the connection. The route established between any pair of connections is not unique. The call finds its own way through the system. Every switch, and virtually the entire equipment, used to set up a call, remains tied down for the duration of the call. It works on a 'step by step' principle, in which each stage involves physical movement of selectors.

The Strowger technique requires very large quantities of switches, each of which is a complicated mechanical device which has to perform fast, jerky motions causing heavy wear and tear. Maintenance requirements, therefore, are considerable (Cripps and Godley, 1978, pp. 31–2).

(2) Common Control Systems – Crossbar

In the 1950s and early 1960s, British development efforts were directed at developing electronic exchanges. These efforts failed. In Britain, the industry was slow to develop (electromechanical) common control systems. In a common control system, like a manual system but unlike Strowger systems, control mechanisms are separate from switches. Once a call is set up, control mechanisms are freed to set up other calls (Cripps and Godley, 1978, p. 32). The British manufacturers recognized the need for a common control system to offer in export markets and ATE (absorbed by Plessey) developed a Crossbar system in the mid-1960s. This has had only limited success in export markets in competition with the systems offered by companies such as Ericsson and Siemens who exploited their ten year lead.

(3) First Generation Electronic Exchanges

By the end of the 1960s, foreign manufacturers were introducing first generation electronic exchanges. These now dominate world markets. During the late 1960s, the Post Office and its suppliers developed the small TXE local exchange which sold to some extent on world markets. The larger TXE 4 was delayed so long that there was little chance of major exports (Cripps and Godley, 1978, p. 22). Manasian suggests that Britain's decline in world export markets has been the result of a series of disastrous decisions by the BPO over the past thirty years (Manasian, 1979).

Crossbar and TXE offer no cheaper initial capital cost per unit of traffic capacity. Their advantages are in terms of less maintenance requirements and better service. By the time of the Carter Report (1977), Japan had Crossbar with programmable control. The United States had solid state switching with programmable control. Programmable control offers advantages of less maintenance, greater versatility and the possibility of offering wider services.

The Post Office had installed two small stored programme control exchanges. But present versions of the TXE 4 have wired in logic. They lack the flexibility inherent in software control. The BPO considered the development of stored programme control for the TXE 4 in the late 1960s, but rejected it on the grounds of cost and the high risk of delay. Wired in logic TXE satisfies the BPO's operational requirements and they claim that the advantages foregone by the absence of stored programme control are not apparent to customers (Carter, 1977, pp. 99-100). However, Manasian has pointed out recently (1979) that most overseas customers are now specifying fully electronic computer controlled exchanges which British firms cannot yet supply.

INDUSTRY STRUCTURE

Three major firms, GEC, Plessey and STC (a subsidiary of the American owned ITT) dominate the supply of telephone equipment to the BPO, accounting for 90 per cent of their purchases. In addition, IBM and Thorn-Ericsson are very important in the supply of new electronic private exchanges (ASTMS paper, 1979).

The Post Office has long sought to maintain a monopoly of the supply of apparatus connected to its network. It maintains that customers' wrongly designed, installed or maintained apparatus could impair the efficiency, performance and even safety of the network as a whole. The BPO insist that close control of design, installation and maintenance are most important for efficient operation (Carter, 1977, p. 376). Allowing outside suppliers to connect apparatus at will would result in outside suppliers 'creaming off' the profitable parts of the business, leaving the BPO to pick up unprofitable business (Carter, 1977, p. 378).

The BPO clings resolutely to the concept of the 'integral network', but has conceded some exceptions. Perhaps the most important economically so far have been large PABXs. A PABX is an exchange which links the private internal system of a company to the external public system. The BPO installs any PABX under 100 lines capacity, but authorizes the installation of larger systems by individual manufacturers (CIS, 1979, p. 21). The BPO licenses thirteen different types of PABX. IBM who entered the market in 1976 with a stored programme control model with data acquisition facilities, has acquired 40 per cent of the market according to CIS. The traditional suppliers to the telephone exchange market are trying to erode IBM's newly established lead. General Telephones and Electronics is introducing a new PABX specially designed for hotels. It offers automatic alarm calls, automatically records and pages for incoming calls and has a table top printer and Visual Display Unit for access to and reproduction of accounts and other records (CIS, 1979, p. 21). In addition, the BPO has allowed very technically advanced American made modems which it could not supply itself to be connected; and automatic burglar or fire alarms and some simple telemetry devices (Carter, 1977, p. 379).

STC admitted that the BPO has 'helped' them to a substantial worldwide market. The BPO had set extremely stringent standards for submarine systems, which STC met subsequently. The reputation for reliability which conforming to BPO standards conferred was a significant factor in the STC success. But often the BPO was considered to be too cautious, slow and rigid. Because of its rigid specifications, the BPO will not accept foreign designed products. Consequently, the products designed to meet BPO specifications and made in the United Kingdom are too expensive to export (Carter, 1977, p. 454). The BPO is concerned with the costs of ownership of equipment through its life. But poorer developing countries may often be more concerned with initial capital cost. Therefore, equipment made to BPO specifications may be too expensive for export markets (Carter, 1977, p. 456).

Cripps and Godley comment that the BPO's rigid specifications for the equipment it is prepared to buy result in equipment that is very reliable, even if technically obsolescent. Despite the fact that as late as 1976 over half the net expansion of capacity was Strowger 'Our telecommunications system, looked at from a user's point of view . . . does not function noticeably worse than that of most other advanced industrial countries' (p. 69). But Cripps and Godley conclude that the BPO have been largely responsible for preventing the supply industry from developing internationally competitive exchanges. This has been an important factor in the relative failure of the supply industry in world markets (pp. 68–9).

However, in our recent study, two manufacturers of equipment to be connected to the BPO network considered that criticism of the BPO was exaggerated. One firm considered that the BPO was no worse than any other national telecommunications authority. The other was full of praise for the BPO. These firms require approval for communications interface. This takes a few months, but this is reasonable and firms' planning can take this necessity into account. On some occasions, one firm had wanted to do something in a hurry and the BPO had reacted very fast with technical instructions – e.g. about the location of isolators, and then given immediate approval for connection to the network.

Post Office annual expenditure on exchange equipment rose very sharply between the late 1950s and 1973–74. Cripps and

Godley calculated that in real terms, expenditure in 1971–75 was more than five times as great as in 1957–60. They comment that

> When growth of demand first started to accelerate, installations did not keep pace and the margin of spare capacity was squeezed. The rate of installation then had to be increased not only to match demand, but also to restore the level of spare capacity. The result was a large 'overshoot' in the rate of installation, which has made the recent cutback sharper than it might otherwise have been. With hindsight, the rate of installation could have been smoothed by building up spare capacity before and during the early stages of the period of fast growth of demand, and then working off excess capacity during later years when increases in demand were at their peak.

Between 1971–72 and 1975–76, Post Office ordering of exchange equipment fell by about 60 per cent. Orders for Strowger equipment fell by about 70 per cent, for Crossbar more than 80 per cent. At the same time, orders for TXE rose by about 40 per cent.

Thus, the effects of the changeover from electromechanical equipment to reed relay semi-electronic equipment, combined with dramatic cutbacks in total ordering, combined to cause widespread redundancies (Cripps and Godley, 1978, p. 18).

COMPETITIVENESS OF THE BRITISH INDUSTRY

The British share of world trade in telecommunications equipment has fallen from 25 per cent in 1963, 10.5 per cent in 1970 to 5.9 per cent in 1977. Import penetration rose from 3.5 per cent in 1967 to 14 per cent in 1976. The major problem has been failure to export to the high growth areas of South America and the Middle East (ASTMS paper, 1979).

The BPO is planning a family of advanced electronic Stored Programme Control (SPC) exchanges based on the concept of integrating transmission and switching by the use of common digital techniques. It involves the progressive introduction, initially as an overlap, of a range of digital switching and transmission modules based on microelectronics and SPC trunk and junction exchanges will become entirely digital (Carter, 1977, p. 101). The exchanges will be modular, enabling them to be configured to suit almost any requirements of BPO or foreign telecommunications authorities. Flexibility is enhanced by the use of highly automated design techniques and computerized information systems which allow design changes and configurations to be made quickly and simply.

Digital techniques, in the form of pulse-code modulation and time-division multiplexing are widely used in the BPO network already. But the system as a whole is still essentially analogue, involving a considerable amount of multiplexing and demultiplexing equipment. System X, by integrating switching and transmission, coupled with SPC operation, will result in a series of exchange designs that are smaller, more reliable in service, more flexible in use and generally superior. Eventually, when System X is brought fully into service, it should provide the basis for better, cheaper telecommunications services. The final result will be a national telecommunications network significantly more advanced than any planned at present (Moralee, 1979a).

FACTORS AFFECTING COMPETITIVENESS

A. Organizational

The Carter Committee considered that the concept of the System X family of SPC electronic exchanges, based on microelectronics, was sound, although ambitious (Carter, 1977, p. 101). But Carter considered that, while it is probably urgent for the suppliers to be able to offer System X on world markets, the availability of System X was not so urgent for the BPO. Further, the Committee considered that GEC was suspicious of co-operation with STC, owned by the American multinational ITT. These firms were all being asked to co-operate in a development project with, at that time, no firm order at the end of it (Carter, 1977, pp. 101–2).

Design, development and manufacturing has been split between three companies. Development work has been split within the companies – for example, Plessey has been developing the digital switch at three different sites. There is also suspicion between the three suppliers. Plessey is developing the exchange switches, but GEC is believed to have an alternative up its sleeve in case the Plessey switch fails. GEC is responsible for developing the computer controls, while Plessey has already developed a computer it considers suitable for System X. STC will be in the difficult position of selling System X against ITT's System 12. ITT is installing its first System 12 exchanges in January 1980, two years before the first System X exchange is installed in Britain and four years before System X is ready for export. Manasian suggests 'should System X falter, access to ITT technology could prove invaluable to the Post Office, if unpopular with Plessey or GEC' (Manasian, 1979).

Manasian pointed out that although development of System X began in 1977, it took until 1979 for the BPO and its suppliers to agree on joint international marketing. Both the Carter Committee and Cripps and Godley criticized the BPO for carrying over its old problems of over-rigid specifications and slow, bureaucratic decision making into the System X project. But Moralee (1979b) suggests that there may now be grounds for optimism that these problems have been largely overcome at last. System X represents a new approach to the whole problem of telecommunication modernization, based on pooling the operational experience of the BPO, the manufacturing experience of GEC, STC and Plessey and the design expertise of the engineering staffs of all four participants. He claims that the three manufacturing firms have 'changed their attitudes and are now working ever-more closely together on a project that involves sharing much previously jealously guarded technological information and knowhow'.

B. Technical

Older generations of equipment will continue to be installed throughout the 1980s. By 1989, System X will still only account for about 15 per cent of capacity. Few of the services to customers made possible by System X will be available to customers until well after 1990 (Manasian, 1979).

Modem-based transmission over the analogue telephone network is inadequate for all but the simplest data communication applications, even using advanced microprocessor based modems offering higher transmission speed (Moralee, 1979a). To meet this need, the BPO is introducing a public Packet Switching Service (PSS).

Packet switching is a technique of directing a stream of data signals into small 'packets', each of which carries a destination address and finds its way through the network. It can be a more efficient means of data transmission than circuit switching. The BPO set up the first network in use outside the United States – initially three experimental packet switching exchanges in London, Manchester and Glasgow, and attracted forty major customers with 150 branches (Carter, 1977).

In 1979, the BPO plans to inaugurate a public packet switching network, incorporating further exchanges in Birmingham, Bristol, Cambridge, Edinburgh, Leeds and Reading. The service is already linked to America and links to Europe are planned soon. The BPO plans to lead Europe in this area (*Financial Times*, 1979; Boothroyd, 1979).

But Moralee suggests (1979a) that 'most observers agree that packet switching services are not necessarily the most suitable for every type of data communication application and that advanced circuit-switched facilities are also needed'. Circuit-switched data services are available in most other countries, but will not be available in Britain until at least the mid-1980s when the System X programme is well under way. He also claims that System X does not take any account of the rapid development of satellite technology, which is already suitable for domestic short links for integrated voice and data traffic. But, such links cater only for high volume users, reducing the resources available for modernization of the terrestrial network for providing basic services to the low volume user. But System X has been designed specifically to cope with the expected rapid rate of change in the demands made on telecommunications networks and in the technologies used (Moralee, 1979b). System X is completely modular. System functions are split between more or less independent modules – ranging from small local exchanges to large main network switching centres co-operating via well-defined standard interfaces. Changing operational demands can be met either by adding modules from the existing range or by developing new ones. Modules can be re-engineered individually to take advantage of new technologies as they become available. Moralee suggests that the modular design of System X 'is seen as giving the range a very real advantage over its more narrowly conceived rivals'.

THE IMPACT OF MICROELECTRONICS

The ever-increasing power and dramatically declining costs of microelectronic components are important in increasing the facilities available in stored programme control exchanges and reducing their costs.

All messages are translated into binary digits in digital systems. With SPC 'the information required to set up and control connections – including dialled systems, customer class of service information, routing and charging information – is stored and manipulated in computer-like data processors in accordance with electrically stored program of instruction' (Harris, 1977/8, p. 16). SPC is very flexible – new services and facilities, and modifications to existing services can be incorporated readily.

Bell introduced their ESS 1 into public service in May 1965. By 1975, 15 per cent of the Bell system was ESS exchanges. In 1976, ESS 4, the first of a new series of giant exchanges, was installed. This has resulted in dramatic reductions in capital cost.

These sytems reduce maintenance requirements by half. Centralized maintenance has doubled maintenance productivity. The measurement of traffic has improved and there has been greater accuracy and accessibility of billing data. Better management and traffic information has made it possible to minimize surplus capacity requirements (Carter, 1977, p. 100).

System X will use microelectronics extensively and this means that 'there will be substantially less equipment to produce, and fewer wires to interconnect'. But the BPO admits that 'System X is unlikely to achieve any very considerable penetration before the mid-1980s and it will not displace all the systems for many years after that' (Harris, 1977–78).

Itemized billing, abbreviated dialling and automatic transfer of calls are already available in the United States. Computer controlled digital exchanges will allow massive transfers of data through the telephone system. They will permit enormous expansion in electronic mail, fund transfer, teletext and facsimile transmission. Connections and disconnections to the system can be made by keyboard instructions rather than by hardware alterations (Manasian, 1979).

REFERENCES

Association of Scientific, Technical and Managerial Staffs, *Some Aspects of Employment in the UK Telecommunications Manufacturing Industry*, 1979.

David Boothroyd, 'PO data transmission tariffs cost computers per packet', *Electronics Times*, April 1979.

C.F. Carter, *Report of the Post Office Review Committee*, Cmnd. 6850, HMSO, 1977.

C.F. Carter, Appendix to the Report of the Post Office Review Committee, Cmnd. 6954, HMSO, 1977.

Counter Information Services, *The New Technology*, CIS, 1979.

Francis Cripps and Wynne Godley, *The Planning of Telecommunications in the United Kingdom*, University of Cambridge, Department of Applied Economics, February 1978.

Financial Times, 'Packet network starts', 23 March 1979.

L.R.F. Harris, 'Introducing System X', *Post Office Telecommunications Journal*, Vol. 29, No. 4, Winter 1977–78.

David Manasian, 'The Unknowns of System X', *Management Today*, August 1979.

Dennis Moralee, 'British Telecommunications and the Information Age', *Electronics and Power*, June 1979a.

Dennis Moralee, 'The System X project', *Electronics and Power*, August 1979b.

12. Televisions

N. Swords-Isherwood

The manufacture of televisions – and particularly colour televisions – accounts for most consumer electronics production in the United Kingdom. After periods of rapid growth following the introduction of black and white and then colour television, British production of consumer electronic equipment has stagnated over recent years. But now, new products such as video recorders and home computers have created the possibility for consumer electronics production to grow relatively fast.

TABLE 12.1 The British electronics industry – major end user segments 1978

(Total value – £2000m)			
Defence Electronics	Computers	Telecoms.	Consumer Electronics
£500m (25%)	£700m (35%)	£450m (22.5%)	£350m (17.5%)

Consumer electronics is an end user segment of the British electronics industry along with defence electronics, computers and telecommunications. Consumer electronics is the smallest. Over the three-year period to 1978 computers grew fastest, followed by defence and telecommunications. Although consumer electronics did not grow over this period, this segment could be the fastest growing in the next five years. Colour televisions account for the largest share of consumer electronics – around three-quarters of total British sales. The share of black and white televisions has fallen from around two-thirds of the industry's sales in 1967 to well under 10 per cent.

HISTORICAL BACKGROUND

Following the introduction of colour transmission in 1967, the two major influences on the growth of the sector in the United Kingdom were technical change and the fluctuations in the market exacerbated by government economic policies. These two factors frequently interacted. The adoption of colour transmission marked the beginning of the decline of monochrome manufacture in the United Kingdom. Relaxation of credit restrictions in the following year increased demand for both types of sets, but British manufacturers were faced with production capacity limits. The consequence was a rise in imports of 700 per cent. The introduction of credit restrictions forced a temporary cut back in demand and this particularly affected monochrome. As demand improved, the domestic supply shortage was met by further imports. In 1969 the change in transmission from 405-line to 625-line made the British compatible with the domestic products of major exporters facilitating further imports. Supply constraints continued for the British manufacturers and imports of both colour and monochrome continued to rise. Japanese exports to the United Kingdom increased dramatically.

Changes in government credit policy and particularly in Value Added Tax exacerbated British manufacturers' supply problems. The 'Barber boom' of 1971 coupled with lifting of credit and rental restrictions marked a major shift of market share from the British manufacturers to imports. Domestic supply problems were compounded by the frequency of changes in VAT rates and their consequence for consumer demand. Hire purchase controls were lifted in 1971, reimposed in 1973 and changed in 1974, 1975 and in 1977. Value Added Tax, applicable to sales and rental charges, changed every year and twice in 1975.

Rental accounts for a major part of sets in use. Rental companies are the main customers for British manufacturers. Changes in the conditions of rental therefore strongly influence the market.

INDUSTRY STRUCTURE

A particular sector such as televisions cannot be adequately defined in terms of enterprises since most firms involved in television manufacture are divisions of conglomerates and multinationals in many cases not primarily concerned with the manufacture of televisions.

Firms in the sector range from large multi-product multinational firms which design and make their own sets, tubes and other components, and which own component and assembly facilities throughout the world, to much smaller firms operating on a national basis and who buy in major components and assemble sets.

The world leaders in the production of the major product in the sector – colour televisions – include five Japanese firms, five European firms and three American firms. Ten companies account for around half the world sales of colour televisions, which totalled 28.4 million sets in 1978. Only one British firm is represented, and it produces 2.3 per cent of world production. Other leading European firms produce 24.8 per cent, leading American firms 21.2 per cent and Japanese firms 48.1 per cent.

Thorn is the largest manufacturer in the United Kingdom and is the only British world leader. Its output is around half a million sets and it accounts for 30 per cent of the British market. The biggest share of the British market is taken by foreign multinationals, in particular the Dutch Philips (20 per cent), the American ITT (9 per cent) and the Japanese firms (10 per cent). These leaders account for almost three quarters of the British market for televisions. There are also some 'up market' foreign companies such as Luxor, Bang and Olufsen and Grundig who maintain sales facilities only, in the United Kingdom. Total British production in 1977 was around 1.6 million sets. GEC and Rank have about 6 per cent each of the market; Rediffusion 5 per cent; Decca and Granada 4 per cent. A variety of other companies take up the remaining 6 per cent (Sciberras *et al.*, 1979).

The Japanese have entered American, then European and British markets. This poses serious challenges to the established industry in the United Kingdom. Licences and import restraints have been factors inducing Japanese firms to set up production facilities in the United Kingdom, both as independent concerns and in joint ventures with British firms. The Japanese entry into British manufacture represents both a competitive challenge and an opportunity to learn about technical changes in products and processes most rapidly and effectively adopted in Japan.

Five Japanese firms now manufacture in the United Kingdom. Matsushita and Sony operate factories as subsidiaries of the parent companies in Japan. Toshiba and Hitachi are involved in joint-ventures with large British firms which own small television manufacturing subsidiaries. Mitsubishi announced the takeover of the British subsidiary of the bankrupt Norwegian firm Tandberg in 1979 and established manufacturing operations.

Other British competitors such as GEC, Rank Radio International and Decca, are all part of large multidivisional companies whose consumer electronics activities form a small part of their total business. Each has only a small share of the British market.

A unique feature of the British television industry is that the largest share of market distribution is accounted for by rental. The percentage of black and white sets rented has fallen from 47 per cent in 1972 to 31 per cent in 1977. Over the same period, the percentage of colour sets rented has fallen from 76 per cent to 66 per cent. Rental outlets account for a major share of the sales of the established firms in the United Kingdom. These are the largest outlets both for the market leaders and for the indigenous firms. The interests of the rental companies have had an effect on the manufacturers.

COMPETITIVENESS OF THE BRITISH INDUSTRY

The competitiveness of the British industry is shown by relative cost comparisons; by comparisons of reliability and quality; by the increased market share taken by new entrants; and by the closure and takeover of British firms.

1. Cost Comparisons

Production costs of a typical set vary greatly as is shown in Table 12.2. Of considerable importance is the question of labour costs. Labour costs are a function of labour time for

TABLE 12.2 Production costs of a typical set (£)

	UK	*W. Germany*	*Japan*	*Korea*
Labour costs	10.6	15.0	5.7	1.5
Material costs	126	119	100	113
Overheads	20	17	11	2
Total	157	151	117	117

Source: E. Williams, 'Time for colour TV makers to look again at restructuring', *The Engineer*, 15 March 1979, p.13.

TABLE 12.3 Manhours and labour costs per hour

	UK	*W. Germany*	*Japan*	*Korea*
Time (in manhours) per set	6.1	3.9	1.9	5.0
Labour costs:				
- per hour	1.74	3.85	3.0	0.3
- per set (£)	10.6	15.0	5.7	1.5

Source: As Table 12.2.

manufacture of a set, and the hourly labour costs. Labour costs per hour in the United Kingdom are lower than those in Japan and West Germany, but time in manhours per set is considerably higher.

Higher British costs are the result of poor performance in design, reduction of components, and production automation. The United Kingdom also has paid less attention to testing and has suffered in product reliability and quality. The disadvantages of a small home market have been exacerbated by excessive proliferation of set types.

Of particular significance in effecting improvements in these areas is reduction in the number of components. While there is no very significant difference between German and British sets, the difference between the European and their Far East competitors is dramatic. Japanese sets have on average only two-thirds of the components of British sets. While the British firms have recently attempted to narrow this difference, the Japanese have gone even further in the use of MSI circuits and developments in the tubes which continue to reduce component content.

2. Reliability and Quality

Quality for consumer products is made up of quality of function, product differentiation and reliability. In the past, European and American firms have stressed the first two of these.

Japanese firms stressed reliability. European and American lack of concern with reliability meant that by the mid-1970s European and American sets were failing at a rate about five times that of Japanese firms. As late as 1977 European and American sets had over twice the breakdown rate of Japanese sets. European and American firms initially refused to see the importance of reliability in competitiveness, focusing on price in the face of declining market share to the Japanese.

The reasons for the different emphasis on reliability are partly historical and partly a result of the structure of competition. Japanese concern with their image for poor quality in the post-war period was a factor. Equally important, however, was the export orientation of Japanese television manufacturers. The high cost of establishing and maintaining a service network in areas as vast as the United States and as diverse in language and structure as Europe meant that Japanese sets needed to be very reliable, in order to make marketing overseas feasible.

3. New Entrants and Plant Closure

Between 1973 and 1974 there was a dramatic trade imbalance in colour televisions. Even though this was rectified by 1977, the new entrants had established their names in the market place and this served as a prelude to yet more new entry.

A comparison of exports in three major European countries and Japan showed that the bulk of British, French and German exports were directed towards each other and other European countries. Germany was the leader of this group, exports being approximately four times those of the United Kingdom. But Japan was the overall clear leader with exports over four times those of the combined European export figure.

An analysis of world markets and production showed that compared with France and Germany, the United Kingdom has both the smallest market and the smallest production of colour and black and white television in value terms. The

United Kingdom is second to Germany in volume except for the production of black and white sets, where France is ahead of both the United Kingdom and Germany (Mackintosh Publications, 1978).

The loss in the British market share has resulted in declining employment and continual closures of manufacturing plant in the United Kingdom. All major manufacturers have announced factory closures and substantial layoffs over the last few years. The United Kingdom has virtually ceased to manufacture monochrome sets, with domestic manufacturers importing sets brought from the Far East for resale.

FACTORS AFFECTING COMPETITIVENESS

The Japanese competitive challenge is the most important threat to the established industry internationally. The United Kingdom is particularly threatened. In the past, protection offered the British industry by the PAL licence has been reinforced by the unique system of marketing, in which the established companies' control over rental outlets has been an effective barrier to entry. These conditions are changing. PAL patent protection is at the point of expiry. This will affect all European competitors.

In the United Kingdom, field maintenance service, the *raison d'être* of the rental business, has become far less attractive as sets are now much more reliable. Consumers' demand for a wider range of products and features has weakened rental's hold over the market. Rental companies are trying to recapture their position by changing their traditionally conservative policy towards new technologies and products. New entrants' shares of the market have grown even in the face of these barriers.

In overseas markets, particularly the United States, these forms of protection have not applied. There, market leaders – including RCA and Zenith – have failed to innovate with competitive new products and have lost significant shares of the market to Japanese firms.

Similarly, in the United Kingdom, the indigenous firms have failed to appreciate the requirements for effective competition early enough. Two 'leagues' operate in the United

Kingdom. While, in the past, there were competive opportunities for little league firms, these have been reduced considerably by the diffusion of new technologies since the early 1970s. Value added in assembly has been reduced dramatically by reductions in the numbers of components in each set and by the automation of assembly and testing. Better quality components have become essential prerequisites for the achievement of set reliability. This has eliminated opportunities for profiting by the purchase of cheap components.

Faced with these difficulties, two of the three indigenous British firms in this category have formed joint ventures to gain access to the product design and manufacturing technology of Japanese big league firms. Neither have the British big league firms succeeded in adopting appropriate competitive strategies.

The Japanese big league firms have adopted automated production techniques most successfully. Non-Japanese competitors have been relatively unsuccessful in production, design, the use and deployment of manpower and marketing. They have not recognized the importance of quality and the price premium that it can yield. The Japanese firms have been more successful than firms in other advanced countries in planning production layout and in integrating product and process design to conform with production and testing equipment. They have been more successful in automating functions such as testing, soldering, conveying and handling. This all contributes to the quality and economy of Japanese automated production. Firms outside Japan have not been as imaginative as the Japanese in their training and manpower deployment policies.

Non-Japanese firms in the United Kingdom have developed pricing strategies aimed at average to low price. Their attitude to marketing remains based, as in the past, on the assumption that consumers prefer cheap products. Japanese firms' prices are based on average to high price strategies. They have succeeded with marketing strategies based on quality and the assumption that consumers are prepared to pay for this. This has given as extra degree of freedom to adopt production techniques which although not dramatically more cost competitive offer advantages. These can be passed on in the form of higher prices.

These factors become increasingly critical as new products,

such as Viewdata and video recorders, are developed which require intimate integration of components and end product. Product design needs to be planned carefully to take full advantage of increasingly capital intensive production processes. For example, it is necessary to ensure that a higher proportion of components can be handled by fewer machines more rapidly and with higher reliability.

Some British government policies have had adverse effects on competitiveness in the industry. Frequent changes in economic policy have caused disruptions to manufacturing plans. The initial high share of imports in colour television demand was encouraged by government relaxation of demand controls at an inappropriate time for British suppliers. This high share was later redressed but this experience did not contribute to positive relations between industry and government.

The declining competitive position of the British industry has resulted in demands for industry rationalization to obtain larger volume production units; and also in the formation of further joint ventures with foreign – particularly Japanese – partners in order to obtain benefits of technology transfer. The government has been asked by NEDO to play a role in encouraging these merger moves.

Failure to improve government-industry relations could be disastrous. Japanese government-industry relations have been close and effective, especially in the development of the Multiplex interactive telephone-television system. The Japanese do not suffer from the traditional American and European 'mistrust' between state and industry and this could prove a critical element in their success in development and diffusion of products crossing boundaries between telecommunications and consumer electronics in the future.

Policies of industry rationalization which fail to recognize these overlaps and which fail to encourage competitive strategies by firms to exploit these will miss important advantages which Japanese competitors have obtained. Mere size is insufficient for competitive success.

Manpower policies and strategies are an area of major weakness in the British television industry. These will be discussed in greater detail elsewhere. For effective competition, major changes must occur. These changes need to be firm-based initially, but in the long term they require more imaginative government support particularly for higher education. Government could also have a direct role in promoting retraining and formulating more positive measures to deal with unemployment. Current cuts in education and training make such policies unlikely in the near future.

THE IMPACT OF MICROELECTRONICS

Technical change has been an immensely important factor in the development of the industry, influencing not only the products' quality, reliability and use, but also the process of manufacture. It has been a crucial influence in the competitiveness of firms and countries. Most changes have been dependent on the development of microelectronics. Technical change has affected the television sector through:

(1) changes in transmission and receiving systems;

(2) changes in the product such as the development of small sets, remote control, hifi sound, flat screens, multiscreen sets, and sets with 'fax' capability. Smaller sets produce heat problems which need new components for size and reliability. Remote control and hifi sound need advanced integrated circuits or microelectronics, as do the more exotic developments;

(3) changes in components and the process of manufacture, in many cases difficult to distinguish from changes in product.

The development of related products and changes in the role of the set have been important, e.g. video games, video cassette recorders and discs. Technical change has enabled the role of the set to change from simple entertainment to providing a home information system for instance Teletext, Oracle and PRESTEL.

Greater capital intensity of production has been made possible by better design and by the dramatic fall in the number of components. Since the early 1970s, the number of components in a typical set has been reduced from over 1,400 to around 400 due to solid state. Consequently far less labour is required for assembly. This has made possible the introduction of automated insertion processes which in turn push this change by giving incentive to reduce component count further. These changes have also dramatically improved reliability. The demand from consumers for greater reliability has influenced the manufacturing process by giving incentives for the use of more rigorous and automated testing procedures.

These technical changes do not operate independently. They interact with each other. Thus, developments in one component encourage or necessitate developments in others. Developments in solid state components are necessary for product changes such as remote control. The changed role of the set also is dependent upon component developments such as decoder chips for Teletext and Viewdata.

Changes in the manufacturing process are dependent upon integration to reduce the number of components to a level 'handleable' by automated machinery. The interdependence of changes has dramatically increased the complexity and importance of manufacturing process design. Improved componentry was vital to these changes but it was a necessary but not sufficient condition for success, which depended to a large extent on a new approach to design capability which the Japanese had.

REFERENCES

Consumer Electronics, Television and Audio Products, Mackintosh Publications, 1978.

E. Sciberras, N. Swords-Isherwood and P. Senker, 'Competition, technical change and manpower in consumer electronic equipment: a study of the UK television industry', Report to the Engineering Industry Training Board, SPRU, November 1979.

E. Williams, 'Time for Colour TV Makers to look again at restructuring', *The Engineer*, 15 March 1979.

13. Minicomputers

N. Swords-Isherwood

Minicomputers are the fastest growing segment of the computer industry. Diffusion into new applications has been rapid. The increased power capacity and price competitiveness of the minicomputer has fostered a change from mainframe computing by blurring distinctions between the two.

HISTORICAL BACKGROUND

Minicomputers were developed about twenty years ago, appearing first in the early 1960s. Limited to 4k or a maximum of 8k words of memory, they were built by electrical engineering firms like the American Westinghouse, the British Farranti, and the German Siemens. The high cost of these early products resulted in them being used mainly in military application and in large power and process automation schemes where quality rather than price was the principal concern.

Minicomputers have had design generations of around five years. With each phase of design development, prices and costs have fallen dramatically, and capacity has increased. These developments have resulted in increasingly diverse applications. The most recent generation of very large minicomputers is compatible with existing minicomputer families but can offer main memory capacities expandable to 1m words. These are equivalent in capacity to mainframe computers but cost little more than 30 per cent of the mainframe competitor.

State market requirements, particularly military requirements, provided the initial stimulus to the creation of markets for minicomputers both in the United States and in the United Kingdom. In both countries, the state continued to stimulate technical change involving increased computer power and smaller computers. Recently, however, industrial market requirements became increasingly important and provided the source of the greatest potential for growth in the industry. This is significant because the competitive strategies required for the newer, faster developing, commercial and industrial markets are entirely different from those required for success in state markets.

INDUSTRY STRUCTURE

Companies that provide minicomputers can be classified into four categories:

(1) pure minicomputer companies that normally derive all their revenues from the minicomputer market;
(2) large general purpose computer manufacturers, which also produce minicomputers and products related to the minicomputer market;
(3) diversified electronics and semiconductor manufacturers, which have expanded into the minicomputer industry;
(4) companies in related fields, which have entered the minicomputer field (Sciberras *et al.*, 1978).

The leading companies in the world's minicomputer market are American multinationals. With the exception of the Japanese market, where Japanese firms are more significant, American multinationals are dominant in the markets of all industrialized countries. In the United Kingdom, Digital Equipment, Data General, Honeywell, Univac, IBM and Hewlett-Packard account for significant market shares. Computer Technology, GEC and Ferranti are the leading British firms. There are many smaller participants such as Logica and Redifon. ICL has recently entered this market.

The dominant force in the industry worldwide has been the American multinationals. These include pure minicomputer manufacturers which have become established in the industry comparatively recently, as well as large general purpose computer manufacturers and diversified electronics manufacturers.

Although the growth of the industry has been dramatic, it has not been smooth. Competition proved fierce and many firms pulled out of the industry. The failures included both giants and small new entrants. Firms like General Electric and Xerox pulled out of manufacture, selling off their computer interests to industry leaders. Generally, among large firms, it was the diversified electronics and engineering firms which failed in the American industry and the new 'pure' minicomputer companies which took the commanding heights of the industry.

These new entrants grew primarily by internal expansion. Very little takeover activity was involved in their growth until, much later, they adopted vertical integration policies into components and peripherals. The horizontal takeovers in the industry were by the general purpose computer manufacturers which, although among the leaders in the minicomputer sector, were clearly lagging behind the two 'pure' minicomputer manufacturers which dominate the industry.

COMPETITIVENESS OF THE BRITISH INDUSTRY

The industry leaders world-wide are all American multinational companies. With the exception of Japan, where national firms hold a more significant market share, these American firms and their subsidiaries occupy the leading positions in all the principal national markets. The British market includes several indigenous British minicomputer manufacturers. These occupy only a small market share but in particular segments of the industry, e.g. military and state markets, they are a more significant element.

FACTORS AFFECTING COMMPETITIVENESS

1. Planning

The need for planning emerged from the requirements of the fast growing market segments which the American firms were developing – the end-user markets. Success in these markets requires the manufacturer to understand users' needs, to choose, develop and design appropriate products and product ranges, and to provide service and maintenance back up, software and even peripheral products (users in the Original Equipment Markets provide these services for themselves). The end-user market is also price competitive.

In such a market environment, manufacturers have to plan their activities in order to achieve their long-run profit objectives. This need was recognized by the American firms and they have developed their planning accordingly.

The British firms do not experience the pressures for planning felt by the American firms in end-user markets. Customers in the market segments which American firms pursue, themselves undertake the activities which customers in the end-user segement require from manufacturers. Thus, the British firms are relieved to a considerable extent of the requirement to plan and provide service, maintenance, software and peripheral support: the experienced state market users provide these themselves. In these state markets, even design of new products involves customers, who tend to specify often quite detailed technical and production requirements. The need to plan product ranges is thus diminished for the British firms.

The ready availability to British firms of government supported markets (including military, educational, central and local government and nationalized industry markets) has tended to be counterproductive insofar as it has tended to reinforce the British firms' lack of commercial competitiveness.

The experience of state market segements has denied the firms the opportunity to develop expertise in those activities essential for success in commercial segments. Not only did they have less experience in providing maintenance support, software support, etc., but their whole marketing orientation was inappropriate. As state customers specify their requirements, British firms had less experience in judging market requirements. This was reinforced by the state customers' emphasis on quality and technology rather than price. The British firms, therefore, failed to appreciate the need for price competitiveness and continued to expect customers to accept products on the basis of their technological 'excellence' rather than on the implications for costs in the short and medium term. End-users who were in activities often far removed from electronics were more likely to be confused than impressed by technology and were concerned with the implications for the total costs of their operations. Service is vital also as down-time involves substantial costs to the business of the end-user. The lack of experience of the British firms severely hampers their credibility in this market segment.

2. Product Standardization

The American firms' emphasis on end-user markets had important implications for their product policies. The American firms were able to specify how customers' needs were to be met. They designed general purpose, standard products to serve general needs. Product plans were aimed to cover a range of general needs with different levels of computing power. The American firms generally reject business which demands deviations from these general specifications if this means changing hardware. But they always offer software and peripheral options. This standardization of hardware meant that firms had to plan their product ranges to cover the widest possible range of needs. This also required planning of product development and upgrading in order not to lose product compatibility. This was to ensure that as customers' computing needs grew, and as new technical developments became available, the firm could meet these needs and offer these products without requiring customers to incur excessive costs of new software and peripherals. Equally important, customers were given strong incentives not to consider competitors' products.

In the absence of clear goals and strategies, British firms' planning had been less effective than their American competitors'. Product introduction often faced conflicting objectives with different echelons of management having different perspectives and thus different expectations from new product lines. British firms had not yet devised clear strategies to upgrade products and to maintain compatibility within product ranges. When new product offerings were made, customers faced difficulties in changeover of software as well as hardware support products. Within the British firms, the lack of planned co-ordination of component supply with end products meant that new development in the two areas did not move in parallel.

The 'arm's length' arrangement posed difficulties through resulting lack of compatibility of products. Also in the development of major new computing products such as the microprocessor, firms lacked a clear perspective on whether the product was to be offered to customers simply as a component or as a computer in its own right. The impact on existing product lines was not taken into account either technically or commercially.

3. Component Supply

The American firms also adopted clear strategies in relation to component supply. The advantages and disadvantages of rapid incorporation of appropriate technical change in components and of in-house versus bought-in component supply were taken into account in their planning.

4. Price

The overall cost advantages of component and peripheral developments and economies of standard production are passed on to customers in the form of low prices relative to computing power. This is a vital element in competitive strategy in this fastest growing but price conscious segment of the minicomputer business. Technical change, product development and production are geared towards the needs of the particular market. The pricing of products is part of a strategy to capture these markets and is based upon a clear understanding of the nature of that market. American firms recognize these needs and plan their strategies accordingly with their profit objectives clearly in mind. The need for price competitiveness places pressures for ever decreasing costs upon the firms, to which they respond in clearly developed corporate strategies.

5. Manpower and Manpower Planning

This is discussed in Chapter 17.

THE IMPACT OF MICROELECTRONICS

The advent of microprocessors (MPUs) has resulted in the development of microcomputers which can be as powerful as minicomputers but much smaller and cheaper. Microprocessors perform arithmetic and logic and are used in conjunction with memory chips for storing operating instructions and data and for controlling input/output functions. Microcomputers are likely to be used increasingly in such dedicated applications as traffic control and electronic point of sale systems. They have already been sold into markets such as insurance brokers and dental practices, where they have been programmed to fulfill a wide range of functions specific to these businesses.

Like the distinction between mainframe computers and minicomputers, the distinction between minicomputers and microcomputers is likely to become increasingly blurred as microcomputers are improved in performance and as minicomputers incorporate MPUs. It has been argued that a whole segment of the minicomputer market could ultimately be served by such devices. However, the leading manufacturers of minicomputers have found that in the five years that the MPU has been available, predictions about the negative impact on the minicomputer industry have not materialized and they argue that the MPU has created new markets in areas where the minicomputer was not an economically viable alternative.

The American firms' introduction of microprocessors contributed to the competitiveness of their products. But microprocessors posed difficulties for British firms. The American firms' size and greater demand for components for minicomputers, gives them the option of manufacturing components on a scale sufficient to achieve the substantial dynamic economies of scale, which are available through 'learning curve' effects in the production of semiconductor components. In-house component production can enable them to plan new component and end-equipment developments to meet market requirements better. Even those American-owned firms without in-house component production facilities can benefit from being able readily to call on advanced component suppliers located near their headquarters.

Nevertheless, both European firms outside the United

Kingdom and Japanese firms have succeeded in developing links with American component suppliers to counter the inherent advantages of American firms. Thus, while British firms had, and still have, some inherent disadvantages in relation to their principal American-owned competitors, the British firms have generally adopted policies which have tended to exacerbate these disadvantages rather than to compensate for them.

For example, lack of planning in the only British firm which had introduced MPUs resulted in a shortage of marketing personnel when the product was offered. Pressure was put on the computer design staff to market the product. This disrupted work in that division which was already over-committed. No clear strategy had been devised regarding the marketing of the product either as a component or as a computing product in its own right.

Falls in the costs of hardware arising mainly from falling costs of microelectronic components have meant that the highest value added comes from systems sales including the provision of software. This has induced further technical change in software as its share of total costs has increased.

REFERENCE

E. Sciberras, N. Swords-Isherwood and P. Senker, *Competition, Technical Change and Manpower in Electronic Capital Equipment: A Study of the UK Minicomputer Industry*, SPRU Occasional Paper No. 8, September 1978.

14. Skill Implications for the Electronics Industries

P. Senker and E. Arnold

The skills required in the various sectors of the electronics industry are tending to converge. Similarly, boundaries between functions such as research, development, engineering and production are tending to become less clear. Indeed, where companies fail to recognize the need to co-ordinate, or even combine, the various functions, their competitiveness can be adversely affected. For example, Chapter 12 on the television industry shows that the success of the Japanese in integrating design and manufacturing plays a significant part in making it possible for them to produce very reliable sets cheaply.

The electronic equipment industries share with other areas of engineering where microelectronics are being applied the need for skills which bridge the interface between mechanical and electronic elements of systems. However, many electronic equipment firms (notably makers of word processors and similar office systems) are essentially assemblers, buying in components and peripheral devices. It is in the peripherals that most of the interfacing takes place, and hence it is the peripheral makers who need the skills for design across the electronic-mechanical interface, while the assemblers tend to need more system design and maintenance skills.

ELECTRONICS ENGINEERS

The main source of high level manpower in this industry is graduate electronics engineers. Even in the United Kingdom, there is general acceptance of the need for university graduates.

There is a great shortage of electronics engineers in the United Kingdom and efforts to use people with other training have not generally been successful. The training of these graduates to use microprocessors is not a problem within the electronics industry and within other industries such as telecommunications which are converging with electronics. These industries cope readily with microprocessors as a natural continuation of their experience with transistors and integrated circuits. Such firms are vociferous and articulate in their demands for electronics engineers.

BPA Marketing and Management wrote up six detailed case studies to assist us in reaching conclusions in relation to the electronics and allied industries. Their findings confirmed ours. All the companies were short of electronics engineers, but none had any problem specifically related to the introduction of microprocessors, even where their previous experience was with analogue electronics. These companies perceived the shift to microprocessors as part of the evolutionary patterns of the past two decades. Their knowledge of logic and systems circuitry has kept up with the introduction of new integrated circuits and microprocessors. Several companies pointed out that many of their staff were so intrigued by microprocessors that they have developed their own skills as hobbies.

There is some resentment in the industry against firms which do not accept their share of training responsibilities, although it is recognized that some loss of trained and experienced manpower to small firms is inevitable. We found several firms which had relied entirely, or almost entirely, on recruiting experienced engineers from other companies.

Smaller firms recruiting experienced design engineers without knowledge of microprocessors said that there was no great problem in training such engineers 'on the job' to design with microprocessors. One small microcomputer manufacturer has employed minicomputer designers without experience of designing with microprocessors. Such engineers are first employed as a training exercise on a mock project of about one month's duration where they acquire sufficient experience of designing with microprocessors. A larger firm also considered that designers competent with transistors and integrated circuits had no problem designing with microprocessors. The first steps in designing with microprocessors came very easily to electronics designers used to less advanced components. This gave them confidence to go on. One week 'hands-on' courses provided by microprocessor manufacturers were thought sufficient. The main concepts designers needed to grasp were those of designing software using more or less standard configurations of hardware. For example, in making a control with microprocessors, it is necessary to design software rather than to add relay contacts. It was emphasized, however, that designers need to understand microprocessor technology.

In the United Kingdom, recovery from the recent recession is exacerbating the shortage of electronics engineers. One company had tried to cope with this by recruiting software people, but found that their lack of knowledge of the hardware was impossible to overcome. Most of the companies were trying to develop salary and career progression schemes to retain their electronics engineers and were inhibited by overall company policies. These measures are clearly palliatives to which companies are forced to resort in face of the continuing shortage of people with the necessary basic education and training in electronics engineering.

Perhaps more important, there is a possibility that the British Post Office's very large System X programme (see Chaper 11) may be extravagant in the use of these scarce skills.

Senior Engineers

There were widespread complaints of a shortage of senior engineers – at project leader level and above – who need to understand products and their interrelationships with the systems of which they form part. It was generally considered that there was an urgent need for high level, up to date, courses to train such people.

TESTING, INSPECTION AND INSTALLATION

Skills required in these functions depend very greatly on factors such as the complexity and design of the equipment involved and on the design of the test equipment. Where electronic equipment replaces electromechanical equipment, procedures tend to become more skill intensive. Semi-skilled, skilled and labouring level effort may often be replaced by technicians.

The experience of one firm we interviewed illustrates some of these problems. They were introducing microprocessor-based equipment to replace electromechanical devices. Within two years, they thought this would be likely to lead to the halving of their installation and commissioning staff. Initially, electronic equipment requires half the labour for installing and commissioning required by electromechanical equipment. In eighteen months' time, when teething problems have been dealt with, this number should be halved again. But the electronic equipment based on microprocessors required some higher level skills, within the installing and commissioning workforce, which were not required before. New skills were required at what the company considered was 'graduate level'. Thus, the company feared that they would be placed in the position where they had to make redundant long-serving employees and take on new people, possibly young graduates, at almost the same time. Also, there was the feeling that the work available, although requiring a high level of skill, might not prove altogether satisfying to graduates. To overcome these problems, a programme of selection from present employees was devised and all employees, including some who had been performing semi-skilled work, were invited to apply. Those passing the selection test – who included both unqualified semi-skilled workers and some with HNCs – were sent on a training course of about seven weeks. This enabled the company to train a small corps of installers capable of dealing with most of the problems of installing the new electronic equipment.

In addition, the firm found that there were more problems involved in commissioning the new equipment – for example, 'patching', i.e. making small program alterations, normally in machine code rather than a high-level language. In order to avoid the commissioning personnel involving Head Office engineers in such day-to-day problems, some of the installers were selected to go on a six-week university course. These people were able to reach the necessary standard for the work specifically required by the company, but may not have reached high enough standards for them to go on to design software.

A television firm suggested that the development of increasingly sophisticated test equipment and changes in product design have deskilled much testing work, but that a minority of testing work still remains very highly skilled. Because of increased component integration, new products have less circuits to test and plug-in modules are used more widely. More of the testing is 'go/no-go' and displays are digital, or give amplitude profiles which indicate defects. There is no problem making contact – the chassis is placed on a unit and the operator pulls a handle to make all contacts. This testing is less skilled than before and is performed by semi-skilled female operators. However, the requirement for highly skilled testers, e.g. for large or complex installations, is as great as before. The problem is that, while previously testers could move from the less skilled to the more skilled testing, the gap between the skill requirements of the two types of testing has now widened.

It was argued by several firms that test and inspection technicians tended not to be geographically mobile. On the industrial estate where one firm operated, technicians tended to move between firms (and to some extent also between occupations) a great deal. In this firm, about 25 per cent of jobs in these technician categories were unfilled. In 1976, a NEDO survey of test technicians found that there was a shortfall amounting to 10 per cent of the numbers employed (Radar and Navigational Aids, 1979, p. 5).

PROGRAMMERS

There is a shortage of experienced programmers capable of designing systems. This requirement appears specific to firms in the 'information sector', where systems are principally electronic. Where microelectronics is used as a relatively small part of a system (as, for example, in temperature control applications), there appears to be less call for this skill.

Diagnostic programmers were regarded as virtually unobtainable by one firm. It was thought possible to convert programmers to diagnostic work, but only 'very special' programmers are considered good enough for this. It was suggested that the lack of diagnostic programmers resulted from the lack of a substantial British computer industry.

DRAUGHTSMEN

One minicomputer firm suggested that draughtsmen were in short supply, and that this had provided an opportunity for draughtsmen to establish bureaux and act as subcontractors. Because of this, the skill shortage could always be overcome, although at some cost. Design draughtsmen were particularly hard to get, detail draughtsmen less so. The shortage of printed circuit board draughtsmen had induced this firm to experiment with computer aided design, using a digitizer tablet linked to one of their own systems. This was expected to provide a good automated aid that could be used by unskilled labour. It could also have sales potential in its own right.

REFERENCE

National Economic Development Office, Radar and Navigational Aids Sector Working Party, *Progress Report 1979*, London, NEDO, 1979.

PART III OFFICES

Introduction

Chapter 15 discusses Office Automation – in particular, word processing – following the pattern established in the industrial sector chapters in Part II. The historical background is followed by brief data and comments on industry structure. Competitiveness of the British industry and factors affecting competitiveness are then considered, prior to discussion of the impact of microelectronics – in particular, the 'electronic office'.

It is not possible at this stage in the development of the technology to identify implications for offices in the engineering industry specifically. Chapter 16 discusses the skill implications of automation in offices in general.

15. Office Automation: Word Processing

E. Arnold

Office automation equipment impinges on the engineering industry and its manpower in two ways: first, the industry produces the equipment; and second, it uses the equipment in its own offices. Word processing has been selected here as a paradigm case within office automation because it is taking off through microelectronics applications, and because it is at the centre of the technical convergence in the information sector: data processing; word processing; and communications. Arguably, the technology resulting from this convergence will be so generally applicable within the information (office) sector, that it will tend to make other techniques marginal.

The bulk of office procedures manipulate symbols – words, letters, numbers – which are imbued with meaning. At the machine level, these are processed analogously: as groups of bits (bytes), each bit-pattern also having meaning. An alternative way to handle information (typically: pictures; music) is to scan an image, break it up into a large number of small points and assign values to the characteristics of each such point. The resulting information is meaningless, except insofar as it can be used to reconstitute or operate in simple ways upon the original image. That is, it is not composed of symbols. This way of representing information is also incomplete: it relies on the imperfect resolving ability of the eye (ear) to overcome the reduction in the quality of the image imposed by sampling it. Where symbols are manipulated, these are not inherently degraded by the process of electronic encoding.

When information can be treated as symbols rather than as images, a number of advantages are gained: it can be represented using fewer bits; it can, therefore, be communicated, processed and stored more cheaply; and more processing is possible because it is intelligible to programs operating at the symbolic level.

In the past, Organization and Methods techniques have been used to improve the efficiency of word processing in offices, but this has largely amounted to changing the division of labour rather than the technology. The typewriter has been unchanged for a century. Introducing word processing functions into the technology itself, instead of leaving the processing entirely in the hands of the office staff, is an important change in the nature of office technology.

There is no commonly agreed definition of 'word processing', not least because manufacturers have an interest in stretching the concept to fit their own products. At its broadest, it is the manipulation, dissemination and storage of textual information. This definition would encompass everything from typing through copying to filing, making 'word processing' almost synonymous with office automation.

IBM have defined word processing as 'the simplification and improvement of traditional typing, secretarial and clerical functions through a managed system of people, procedures and modern office equipment' (McLean and Rush, 1978, p. 41). Even this is rather broad, reflecting IBM's strategy of selling an 'Office System'. Most manufacturers would want to limit the term to the automation of typing functions, and that is how it is used in this chapter.

The general importance of microelectronics is the quantum reduction it offers in the cost of processing information. In general, tasks can be split into 'thinking' and 'doing'. Where the 'thinking' can be reduced to a set of instructions or rules it has for some time been feasible to automate it, using electronics. Microelectronics increasingly makes it economic to do so, as components are packed more densely on chips and the cost per component therefore declines. However, microelectronics does not merely allow the proportion of system functions which are electronic to be expanded. It also increases the number of functions which the system can economically encompass. In word processing this means that machines are not only increasingly electronics-based, but also offer more processing power: more functions.

HISTORICAL BACKGROUND

Word processing, that is storage media controlled typewriting, has a history stretching back before the 1930s, when the two best-known producers were Schultz and Hooven. Paper tapes began to be used as a storage medium in the 1940s, permitting sequential editing of text. In 1964, IBM launched the Magnetic Tape Selectric Typewriter, offering magnetic, rather than paper, tape and effectively doubling the packing density, but providing no other substantial improvements on existing techniques. The Magnetic Card Selectric Typewriter was launched by IBM in 1969, and while this offered individual page-sized chunks of storage medium, editing and resequencing capabilities were very limited.

In 1971–72, Lexitron and Vydec offered screen-based word processors, incorporating significant amounts of electronic memory and thereby greatly increasing editing power. IBM then launched the Magnetic Card 82 Typewriter, with the same card storage capacity as before, but also 8,000 characters of electronic memory, permitting extensive editing. Later enhancements included some automatic lay-out aids, but IBM still offered no screen-based system and hence no simultaneity of keyboarding and printing. By this time, however, IBM's hold on the American office equipment market was already considerable. In 1977 it was estimated that 80 per cent of installed word processors in the United States were IBM (Wharton, 1977).

Most manufacturers currently offer floppy disks as the magnetic storage medium, and word processors may broadly be classified as belonging to one of three groups:

(1) memory typewriters – typewriter-based systems, usually with fairly small storage capacities, using media other than floppy disks;

(2) stand-alone word processor systems – usually screen-based, with floppy disk storage and simultaneity of input and output;

(3) shared logic systems – screen-based input work stations with access to common processing, storage and printing hardware.

The distinction between these categories is not always clear cut. A position intermediate between stand-alone and shared logic systems is also emerging, where stand-alone systems can be clustered and intercommunicate. In some cases, this permits printing or additional storage devices to be used in common. The clustered approach allows users to experiment with word processing and to expand incrementally without a large prior investment in common elements. However, the marginal cost of adding a stand-alone to a cluster is considerably higher than adding an extra workstation to a shared logic system.

John Lloyd (1979) suggests that stand-alone word processors are economic where up to six workstations are required, and shared logic is economic in the range six to forty workstations. Where more than forty workstations are required, these are most economically driven by a mainframe computer.

Most stand-alone word processors offer a communications option, and many can be interfaced with optical character recognition (OCR) machines, typesetting equipment and mainframe computers. Increasingly, word processor manufacturers are offering limited calculating abilities for their products, while more data processing orientated products are being offered with text processing software.

This is symptomatic of an increasing technical convergence within the three information technologies: data processing; word processing; and communications. Not only can different types of information equipment increasingly be connected together, but in certain cases equipment can be used in alternative roles. For example, one company makes a data capture minicomputer and a shared logic word processor using the same central processor and mass storage hardware, but different terminals and software.

INDUSTRY STRUCTURE

Most word processor manufacturers are assemblers: they buy in standard components and peripherals and assemble them, rather than designing and producing these themselves. This imposes constraints on product differentiation through design and development in that assemblers have no control over the design characteristics of components available to them. Thus, for example, specifically text-oriented software has to be written to run on general-purpose microelectronic hardware. At present, most word processors are based on the industry standard microprocessors: Intel 8080 and Zilog Z80. This pattern could change if component makers decided to diversify downstream into word processing, as Texas Instruments did very successfully with pocket calculators and as Intel, Fairchild and Texas Instruments are currently doing with minicomputers. As more and more of the system consists of microelectronic components, the share of value-added remaining with assemblers falls. However, acquiring electro-mechanical skills would be an obstacle to diversification by component makers.

At present, there are forty-nine suppliers of word processors in the United Kingdom market (Schoeters, 1979). These range from large companies with integrated product ranges extending beyond word processing, through suppliers of word processing only, to small firms with varying degrees of financial backing and product specialization. The slower the market growth, the higher the likely attrition rate, particularly among the smaller less capitalized firms. Recent failures include Wordwright and Dataplex.

COMPETITIVENESS OF THE BRITISH INDUSTRY

A perception of the type of competitive strengths in the industry was instrumental in establishing the National Enterprise Board office systems company: Nexos. Given the negligible British penetration of its own market, only the state has been able to establish a concern with access to sufficient resources to enter the big league. Nexos' approach – federating independent producers and marketing compatible products – offers the possibility of an adequate response to the breadth of IBM's range. Given the independence of the producers within the Nexos federated structure, their desire to make their own products viable in their own right could be turned to advantage in that individual producers' interests are less likely to be subordinated to those of the federated structure. However, Nexos' present thrust at the upper end of the market through Logica VTS and Muirhead needs to be supplemented at the lower end by memory typewriters and similar products in order to offer incremental transition for users into word processing, as opposed to the vast technological leap into the dark at the upper end of the market. In view of the virtual disappearance of the domestic typewriter industry, an attack by Nexos on this lower end of the market could be difficult, particularly as there would be no established user base on which to build. One way to help strengthen the indigenous capability in word processing might be through government procurement policy favourable to British firms.

The market for word processing and related office automation equipment in the United Kingdom is very new, as is reflected by the small number of word processors in use. Correspondingly, there are few relevant statistics available which can be analysed for trends.

An impression of the increasing interest in word processing may be gained from the attendance figures at the Business Equipment Trade Association's annual word processing exhibition in London: see Table 15.1. One firm which had exhibited

Table 15.1 BETA exhibition attendance 1976–79

Year	Attendance
1976	2,593
1977	4,470
1978	6,233
1979	10,014

Source: BETA.

at these exhibitions said that the level of awareness of word processing among visitors to its stand was markedly higher in 1979 than previously. Typically, prospective purchasers had checklists of machine features to work through, and visitors were no longer asking 'What is a word processor?'.

British overseas trade statistics for office machinery can be examined in order to gain some understanding of the industrial background to word processing. In considering trade statistics, it is important to note that they provide no guide either to the ownership of firms or to who has control of the technologies involved. Thus, it is not possible to infer from a healthy trade surplus in a high technology commodity that there is good indigenous technical capability for the production of that commodity, or that the ownership of the means of production are domestically controlled. Where such surpluses occur as a result of the activities of multinational company subsidiaries, the coincidence of interest between the multinational involved and the state is accidental.

The British trade statistics relating to SITC category 714 (Office Machinery) for the three years 1975–77 were reclassified so as to provide, as far as possible, compatible categories across the three years, and are reproduced in summary form below. In addition to calculating trade balances, an effort was made to compare the relative levels of technical sophistication of imports and exports by using value to weight ratios (VWRs) as a crude measure.

Generally, the more technologically sophisticated and therefore 'up-market' a product is, the higher is the ratio of its value to its weight (VWR) compared with other goods in the same category. The export and import VWRs can simply be compared by dividing the former by the latter. Thus, values above unity imply greater technical sophistication of exports than imports. Values below unity imply the reverse. (see Table 15.2). A weakness of this type of analysis is that VWRs can be affected by short-term variations in exchange rates.

For 1977, the only cells where export VWRs were greater than import VWRs were non-electronic calculators and address plates, technologies with a fairly lengthy past and very little future. 'Automatic typewriters controlled by data media' were shown separately for the first time in 1977, when imports were £11,353,000 and exports were £5,463,000, giving a negative balance of £5,890,000. Import and export VWRs were identical, suggesting that exports were re-exports. This inference is corroborated by the low British share of its domestic word processor market. The NEDO Office Machinery Sector Working Party remarked in its March 1979 report: 'Not much more than 5% of current UK placements have been supplied from UK sector firms undertaking some word processor manufacture, and even here, a substantial proportion of components incorporated into their equipment was imported.' (NEDO, 1979, p. 7).

Estimates of the number of word processors installed in the United Kingdom at the end of 1978 tend to be in the range 9-11,000 (industry discussants and press). The National Computing Centre put the figure at 10,720 (Large, 1979). Sales are generally expected to rise by about 30 per cent per annum

TABLE 15.2 British Trade Balances and Value to Weight Ratios, Office Machinery (SITC 714) 1975–1977

	1975			
	Imports £000s	*Exports* £000s	*Balance* £000s	*Export/Import VWR*
714 Office Machinery	479,337	429,989	–49,348	–
714.1 Typewriters, other than typewriters incorporating calculating mechanisms; cheque writing machines	26,916	15,469	–11,447	0.54
714.2 Calculating m/c; accounting m/c, cash registers, postage-franking m/c, ticket-issuing m/c & similar m/c, incorporating a calculating device	42,596	36,467	–6,129	0.57
714.3 Automatic data processing m/c and units thereof; magnetic or optical readers, m/c for transscribing data onto data media in coded form, and m/c for processing such data, not elsewhere specified or included	202,451	212,052	+9,601	0.90
714.91 Other office m/c (e.g. hectograph or stencil duplicating m/c, addressing m/c, coin-sorting m/c, coin-counting and wrapping m/c, pencil sharpening m/c, perforating and stapling m/c)	7,300	23,103	+15,803	0.69
714.92 Parts and accessories (other than covers, carrying cases and the like) suitable for use solely or principally with m/c of a kind falling within 714.1–714.91	200,074	142,898	–57,176	0.61

Source: Calculated from Overseas Trade Statistics of the United Kingdom, 1975, 1976 and 1977, HMSO, London.

1976				*1977*			
Imports	*Exports*	*Balance*	*Export / Import*	*Imports*	*Exports*	*Balance*	*Export / Import*
£000s	*£000s*	*£000s*	*VWR*	*£000s*	*£000s*	*£000s*	*VWR*
681,189	515,531	–165,658	–	818,694	628,027	–190,667	–
43,480	21,488	–21,992	0.55	47,121	25,929	–21,192	0.60
54,721	37,948	–16,773	0.64	47,327	45,844	–1,483	0.82
333,895	272,403	–61,492	0.78	462,251	340,633	–121,618	0.85
9,794	29,355	+19,561	0.68	11,869	33,270	+21,401	0.70
239,299	154,337	–84,962	0.44	250,126	182,351	–67,775	0.53

during the next few years (industry discussions; Hugo, 1979, p. 106; Schoeters, 1979) in Europe.

At present, word processors are novel and relatively expensive. Large and multinational firms are generally seen as the most important potential customers. Expanding the market beyond such firms involves overcoming the market resistance to a new product. The National Computing Centre suggests that by 1987 the price of a communicating word processor will have fallen by 70-80 per cent (Large, 1979).

FACTORS AFFECTING COMPETITIVENESS

The proliferation of models indicates a lack of consensus about what should constitute a word processor, or what distinct bands of capability should be offered within the market for word processing. As a body of experience is built up among users, a clearer pattern of demand is likely to emerge. Again, the sequence which operators must go through in order to execute specific operations is not standardized between machines. This reflects the debate in the industry about the optimum number of keys with which the operator should be presented. This debate between many single-function keys and few multi-function keys to be pressed in sequence is also likely to be resolved through some consensus within the industry. However, product differentiation in hardware is limited by the fact that word processing manufacturers are principally assemblers of bought-in components. It follows that other aspects of product differentiation, such as software, will be important as will company attributes – particularly in terms of whether the firm is regarded as solid enough to stay in business, and the quality and price of maintenance.

Mueller and Tilton have developed a four-part classification of the stages through which an industry passes: 'innovation, imitation, technological competition and standardization' (Mueller and Tilton, 1969, p. 571). They stress that imitator firms try to improve on the original innovation, rather than merely copy it (p. 574). Research and Development costs tend to become a barrier to entry at the technological competition stage (pp. 576–9). For word processing, the technological competition stage is likely to involve considerable efforts in software and many suppliers are likely to be unable to stay in the race as expensive software must be amortized over an increasing sales volume. The achievement of higher component densities in microelectronics will increase the scope for powerful software at marginal extra hardware cost.

As products become increasingly standardized, high volume becomes important to firms wishing to remain price-competitive. Sciberras (1977) has shown how in microelectronics this has led to the emergence of two firm types: 'big league', engaging in price competition and producing highly standardized products; and 'little league', smaller firms who are unable to compete in the volume markets, but who can take greater unit profits across low-volume 'specials' production. A similar pattern could emerge for word processing, as small firms innovate into the gaps between the standardized products.

Product strategies divide neatly between IBM and the rest: IBM is maintaining its 'Office System' approach, trying to offer a complete range of automated aids for the office on the basis of a vast customer base. As one industry discussant pointed out, IBM do not offer 'full frontal' competition to up-market, screen-based word processor systems, although IBM equipment can be configured to operate in a similar way to these word processors. Effectively, they are making a first clumsy stab at the electronic office. Other manufacturers offer less breadth of scope.

In the final analysis, the shape of word processors in a few years time will be significantly influenced by the interests of the survivor firms. Inevitably, IBM must be a force to be reckoned with. Important competitive advantages are secured by:

a wide product range;
compatible or modular products;
an established base of customers;
a 'solid' image, suggesting that the firm is likely to remain in business;
good service and maintenance networks;
economies of scale, both in manufacture and service.

Given its 80 per cent of American word processor installations (Wharton, 1977), and its strength both in the electric typewriter and mainframe computer markets, IBM has a head start with any office equipment it offers.

THE IMPACT OF MICROELECTRONICS

Microelectronics, by making word processing etc. economically viable, has brought nearer the advent of the 'electronic office'. A number of commentators (Barron and Curnow, 1978; Coulouris, 1977; Page, 1980) argue that developments in office technology will tend to the almost total displacement of paper as a storage and communications medium from offices. They look to the provision of an information 'ring main' in offices, (Barron and Curnow, 1978, p. 24), an 'electronic desk top', (Page, 1980), and so on. Such visions depend for their realization upon widespread ownership of relevant capital equipment and on establishing adequate standards and infrastructure for its interconnection. An important constraint will be the need for communication between the 'haves' and the 'have-nots' of office automation, particularly communication between offices and individual citizens. Paper remains the only effective medium for communicating over a distance with someone who has not made any investment in receiving equipment. In the long term, then, the form of non-paper communication will be critically dependent upon the type of equipment generally in use, both in offices and in homes. In the latter case, Viewdata systems (such as the Post Office's PRESTEL) would seem to be technically strong contenders – particularly if the message-sending capability originally specified for PRESTEL were implemented – although the launching of PRESTEL in the United Kingdom has been less than auspicious.

The completely electronic office, if it does materialize, is a fairly long-term prospect. Probably, the electronic office is best seen as the logical end-point of an evolutionary process. Factors affecting that process can be grouped as: technical, market and social. Of these, the social may be the hardest to predict, involving as they do people's responses to change. The rate of employment in the economy as a whole colours people's attitude to labour-displacing technology. Also, the degree of trade unionization of the labour force is important. The print unions, for example, have delayed word processor-based typesetting with considerable success. Trade union strength has, however, tended to be less important in areas of predominantly female employment, such as offices, with the union movement seeing its task primarily as the protection of male employment.

Technical and market factors affecting the importance of word processors in office automation are bound up with their role in electronic as well as paper forms of communication. The price of microcircuitry has considerable bearing on the economics of telecommunications between word processors. Barron and Curnow state that 'communication is inherently expensive in comparison with processing' (Barron and Curnow, 1978, p. 65). Telecommunications switching of information, notably packet switching, involves passing information through a number of switches, each of which is of a considerable order of complexity. Economies of scale and change in switching technology reduce the impact of this complexity on price. But, while the cost of processing continues to fall, that of telecommunications will tend to stay constant or rise. Hence, instead of the current pattern where most data transmission is generated by access to remote processing facilities, the bulk of data traffic will increasingly be for 'Access to remote services which it is uneconomic to provide locally; [and] Transfer of information' (Barron and Curnow, 1978, p. 66).

For word processors, this implies that telecommunications will tend to consist of document-level transfers (including forms of electronic mail) rather than of transfers of information as part of processing. This is distinct from the type of *intra*-office inter-word-processor communication facilities needed, which will include the need to access information as part of the processing itself.

So there appear to be reasons why communicating word processors may prove more important than facsimile in electronic mail, particularly since word processors assist in the composition of messages to be transmitted and can generate messages without a paper original. But, communicating word processors suffer a number of disadvantages too, compared with facsimile and conventional mail. These include authentication of messages via letterhead and signature, which cannot be transmitted via a word processor, and the fact that a facsimile transmitter/receiver, since it is capable of doing less than a word processor, is inherently cheaper.

In developing a model of user choice for message services, Communications Studies and Planning Ltd. (1978) found that there were many important variables in the decision between message services. They found that users tended to use one form of message service quite intensively, and others little or not at all, a factor likely to reinforce institutional conservatism in the form of message sending used. The problem of authentication associated with message services extends to filing based on electronic word processor output rather than on hard copy of some form (paper, microprint, etc.).

While inter-word processor communication may be important in the longer term, institutional factors combined with the low penetration of word processing hitherto suggest that the use of word processors for inter-institution communication may be problematic in the short term, and such devices are likely to find greater application in *intra*-institution communication.

These considerations suggest that the considerable amount of criticism of the Post Office for its slowness in extending data transmission facilities over the packet switched telephone network (PSTN) and introducing a packet switched data network (PSDN) may be premature, at least as far as communication between word processors is concerned. One word processor maker visited was very much of this opinion. The adoption of the internationally agreed X25 data transmission and signalling protocol for the Post Office PSDN will facilitate inter-word processing communication, though there is no formal or *de facto* standard as yet established at the level of word processor software, as opposed to the transmission and signalling protocol, which permits different manufacturers' word processors to intercommunicate, and this could prove an additional obstacle.

The Post Office estimates that it will have only 4,700 PSDN users by 1983, but the *Report of the National Committee on Computer Networks* (1978) argued that 13,000 would be a better estimate (Moralee, 1979, p. 438). The NCCN points out that 'it would not be an exaggeration to say that up to 30 per cent of current mail items are candidates for electronic transmission' (NCCN, 1978, p. 73). Further, the Carter Report points out that mail services are inherently labour-intensive and prices will, therefore, tend to inflate faster than the Retail Price Index (Carter, 1977), making capital-intensive electronic mail more attractive.

REFERENCES

Iann Barron and Ray Curnow, *The Future of Information Technology*, Report to the Computer, Systems and Electronics Requirements Board of the Department of Industry, Science Policy Research Unit, (mimeo), 1978; now published as *The Future with Microelectronics*, London, Frances Pinter, 1979.

C.F. Carter, FBA, *Report of the Post Office Review Committee*, (Cmnd. 6850) and *Appendix* (Cmnd. 6954), HMSO, London 1977.

Communications Studies and Planning Ltd., *A Model of User Choice for Message Services*, Ian Young and John Moulton, London, CSP, October 1978.

George Coulouris, Ian Page and Tony Walsby, *The Potential for Integrated Office Information Systems*, London, Queen Mary College and Butler Cox Partners Ltd., December 1977.

I. St. John Hugo, *Guide to Word Processing Systems*, London, Computer Guides, 1979.

Peter Large, 'The Year of Computerised Secretaries', *The Guardian*, 2 May 1979.

John Lloyd, 'Options Open to the Managers', *Financial Times*, 10 July 1979.

J.M. McLean and H.J. Rush, *The Impact of Microelectronics on the UK: A Suggested Classification and Illustrative Case Studies*, SPRU Occasional Paper Series No. 7, June 1978.

Dennis Moralee, 'British Telecommunications and the Information Age', *Electronics and Power*, June 1979.

Dennis C. Mueller and John E. Tilton, 'Research and Development Costs as a Barrier to Entry', *Canadian Journal of Economics*, Vol. 2 (No. 4), 1969.

National Committee on Computer Networks, *Report of the National Committee on Computer Networks*, London, HMSO, October 1978.

National Economic Development Office, Office Machinery Sector Working Party, *Progress Report 1979*, NEDC, London, 1979.

I. Page, *Office Automation*, State of the Art Report, Maidenhead, Infotech International, February 1980.

Ted Schoeters, 'A View from the Keyboard', 'Paper May Go Out of Date', *Financial Times*, 10 July 1979.

Edmond Sciberras, *Multinational Electronics Companies and National Economic Policies*, Contemporary Studies in Economic and Financial Analysis, Greenwich, Conn., JAI Press, 1977.

Keith Wharton, 'Special Report: Word Processing', *The Guardian*, 9 May 1977.

16. The Impact of Office Automation on Skills

E. Arnold

The productivity of manual workers was increased through automation by some 83 per cent between 1960 and 1970, compared with 4 per cent for office workers, according to Tom Stewart (1979, pp.29–32). There are two reasons for this: first, there was very little technology available in that period which could economically automate the bulk of office information handling; and, second, the division of labour in offices tends not to occur along lines analogous to those in shopfloor production.

ORGANIZATIONAL FACTORS

The division of labour in production has been analysed by the classical economists, notably Smith, who used the famous example of a pin factory to show that:

> The great increase in the quantity of work, which, in consequence of the division of labour, the same number of people are capable of performing, is owing to three different circumstances: first, to the increase of dexterity in every particular workman; second, to the saving of time, which is commonly lost in passing from one species of work to another; and, lastly, to the invention of a great number of machines which facilitate and abridge labour, and enable one man to do the work of many. (Smith, edited by A. Skinner, 1974, p. 112).

Charles Babbage – best known for his pioneering work on calculating engines, and often regarded as the 'father of the computer' – was so impressed by Smith's analysis, that he investigated the process of pin making in some detail. He found that a further advantage resulted from the division of labour, namely:

That the master manufacturer, by dividing the work to be executed into different processes, each one requiring different degrees of skill and force, can purchase exactly that precise quantity of both which is necessary for each process; whereas, if the whole work were to be executed by one workman, that person must possess sufficient skill to perform the most difficult, and sufficient strength to execute the most laborious, of the operations into which the work is divided. (Babbage, 1835, pp. 175–6; italicized in the original.)

Thus, in general, advantages are gained in production by dividing labour, increasing the economic efficiency of individual operations, and cheapening those operations both through this increased efficiency and by applying the appropriate 'grade' of labour to each detail task. The larger the output to be produced, the greater the advantages of division of labour, because labour can be increasingly finely divided. Thus, as Smith has it, 'the division of labour is limited by the extent of the market' (Smith, ed. A. Skinner 1974, p. 121).

This classical economic analysis of the division of labour is not entirely appropriate to office work concerned with text manipulation. Here, the paradigm case (to which there are many exceptions) is the boss–secretary division. Rather than an exhaustive division of labour within the office, as there is within the factory, there is a divison of labour among the bosses which is reflected in the office organization chart. Secretarial services, however, are not generally provided by a specialist team operating as a managed category of divided labour and appearing as a vertically bounded division of the hierarchical organization chart. Instead, they are provided by people who are specifically attached to one or more authors, are 'managed' by those authors, and derive their status from them rather than from their own competence. Peter Gorb argues that 'We have transferred domestic service from the home where we can't afford it to the office where it is deductable for tax, and where secretaries make tea and porters and messengers take on the role of footmen' (Gorb, 1979). That the secretary should be seen as being in a crypto-servant role is evident from the range of tasks expected of this one category of individual in systems where labour is otherwise highly divided. In general, secretaries spend a maximum of only 20 per cent of their time typing (industry discussions.)

Some writers (Oakley, 1976; Benet, 1972) argue that this crypto-servant role – that of 'office wife' – is responsible for the fact that a high proportion of secretaries are women. The 1971 Census of Economic Activity (1975) showed that 98.8 per cent of British 'Typists, shorthand writers, secretaries' were women.

The increasing tendency towards the centralized provision of secretarial and other services in offices represents a shift towards the type of division of labour found on the shopfloor. It presents both new problems of management, and offers an environment within which productivity-enhancing innovations (such as word processing equipment) offer meaningful advantages. As long as a secretary's time is at the disposal of a boss, productivity is limited by that of the boss. Thus, if a secretary's typewriter is used only 20 per cent of the time to process the boss's textual output, then a word processor bought to replace it might reduce that proportion, but the organizational structure (boss–secretary) would not permit the savings which result to accrue to the benefit of the firm or administrative unit. The principal effect would be to raise the cost of doing the same amount of work. Where secretarial services are centralized, however, a classical division of labour can be applied and technical changes can enhance productivity.

Office automation, therefore, implies not only technical but also organizational changes in offices. At the level of the text originator, resistance to office automation may be expected not merely because of simple conservatism but also because the organizational changes involve the loss of the material advantages offered by the crypto-servant secretary. Clearly, where services have already been centralized, automation will involve less resistance on the bosses' part.

TECHNICAL FACTORS

Word processing, as part of a pattern of increasing automation in offices, can directly affect some three-quarters of a million secretaries and typists (1971 Census data). In 1977, there were 81,910 in the engineering industry (*Department of Employment Gazette*, July 1978). The number of people likely to be affected indirectly is harder to estimate. Clearly, word processing has the potential for subsuming other fuctions, of which perhaps the most publicized is typesetting, thus displacing certain other categories of skilled and unskilled labour. In the longer term, message-sending and filing are likely to be affected.

This broader pattern of office automation will be related to cheaper computer power, and can encompass virtually all clerical functions. Some word processing equipment can already store standard form formats, allowing form-filling to be electronically simulated. Other types of electronic data capture can involve encoding via menu selection or by keying pre-coded information from, for example, punching documents. In fact, keyboard-based data capture systems are the closest technical analogue to word processing systems, and in a number of cases, the same equipment can perform both data capture and word processing functions. The market for this type of data capture equipment (key-to-disk) is thought likely to increase in Europe at about 15 per cent per annum, but slightly slower in the United Kingdom. Optical mark recognition is already established for data input, but it tends to be both inconvenient and error-prone. OCR, using special type fonts, is also established – though largely as an input medium for word processing – but is not in widespread use. More direct forms of input (handwriting, speech) are still very crude, and considerably more sophisticated artificial intelligence models will be needed before these are of use in any but the most restricted applications. For some time to come, therefore, data input to, and interaction with, electronic systems will remain a specialized task tied to a specialized input device which must be *operated* and the operation of which will require particular skills – generally a keyboard.

EFFECTS ON JOBS:

Numbers

Estimates of the effects of new office technology on employment abound, but assumptions tend not to be made explicit. APEX suggests that 250,000 office jobs will be lost in the United Kingdom by 1983 (*Electronics Times*, 1979). The French 'Nora Report' (1978) suggests that new technology could displace 30 per cent of banking and insurance staffs in the next ten years. ASTMS are cited as forecasting a loss of 2.6 million jobs by 1985 (Barker and Downing, 1979, p. 17).

The following examples of the effects of word processing implementation on employment were found in published sources:

a major bank has cut typing personnel in one division from 154 to 49 (Schoeters, 1979).

Bradford Council reduced its Jacobs Well staff from 44 to 22 with the introduction of 9 word processors, resulting in an increased productivity of 19% and an estimated annual saving of £58,000–£59,000.

The Provident Financial Group in Bradford had reduced their full-time staff from 27 to 17 and part-time staff from 13 to 3; they had increased their workload; the staff had been reduced through natural wastage; there was a monitoring of productivity but no incentives were paid.

The British Standards Institute introduced word processors and reduced their secretaries and typists by a third.

The CEGB reduced their 'girls' from 50 plus to 25.

The NCB Mining and Research Establishment, Bretby, are considering installing word processors possibly to reduce their typists from 35 to 26.

NCB Western Area have installed word processing on a pilot basis and have solved their staffing problem by 'natural wastage' reducing it from 20 to 14. (Barker and Downing, 1979, p. 18).

'A Nationalised Industry Office' – staff reduced from 20 to 14.

Halifax Building Society – workload trebled with existing staff unchanged.

Civil Service (Pilot Project) – shared logic system. Productivity improvements fluctuating between 20 per cent and 70 per cent of control group.

Large Pharmaceutical Company – shared logic system. Productivity improvements of 127 per cent achieved.

A Large Management Consultants – using varied equipment, measured an increase of 25 per cent in workload with a reduction in staff from 80 to 39 (APEX, 1979, p. 16).

However, the usefulness of such forecasts and data may be more polemic than predictive. A number of factors will affect the amount of labour displaced by word processing. These include:

the extent to which 'best practice' word processing is used and how much it displaces 'best practice' typing methods;

the extent to which the greater productivity of word processing and the high quality of word processed output are inducements to increase the typing workload;

the size of the offices into which it is introduced and, therefore, the extent to which economies of scale can be reaped through word processing;

the price of word processing equipment, both in absolute terms and relative to the price of typing labour;

the extent of worker and trade union resistance;

the 'work mix'.

More generally, the competitive situation of the industries able to use word processing will affect the rate of adoption.

A fairly widespread, and often expensive, discovery made by new computer users during the 1960s was that new information technology alone does not solve problems. Computerizing a disorganized system leads to computerized disorganization. The same effect can occur with word processing and the effects of processing on productivity depend on the extent to which it is incorporated into an organized, managed system. (The shortage of 'applications analysts' able to devise orderly systems into which word processing may be put should be noted.) Thus, the US General Accounting Office found recently that within the American civil and military services, only the Army were making cost-effective use of word processing. The Inland Revenue were using some of its machines for as little as 1½ hours per day (*The Economist*, 1979, p. 95).

Up-to-date statistics on the numbers of people employed as secretaries and typists are difficult to obtain, as official statistics tend to be collected on an industry rather than an occupational basis. Where occupational statistics are available, the level of aggregation tends to be too great for the present purpose.

The Department of Employment publishes a useful breakdown of occupations within the engineering industry. However, for the whole economy, the most recent occupational breakdown at this level of disaggregation is that of the 1971 Census (10 per cent sample).

This implies a population of about three-quarters of a million

TABLE 16.1 Secretarial and clerical occupations in engineering, May 1977

	Total	*Men*	*Women*
Office supervisors	30,190	19,620	10,570
Secretaries, typists	81,910	940	80,970
Office machine operators	39,890	3,940	35,950
Clerks, reception, other	223,640	98,600	125,040
Total	375,630	123,100	252,530

Source: *Department of Employment Gazette*, July 1978.

TABLE 16.2 Clerical Workers, 1971 Census (10 per cent sample data)

	Total	*Men*	*Women*
Office managers	7,382	6,286	1,096
Clerks, cashiers	242,074	90,108	151,966
Office machine operators	16,770	2,211	14,559
Typists, shorthand writers, secretaries	76,794	921	75,873
Civil service executive officers	6,191	4,742	1,449
Total	349,211	104,268	244,943

shorthand writers, secretaries and typists in 1971. Most were women. Estimates of total office employment should be seen in this light. Thus, Wharton says in his word processing 'Special Report' in 1978: 'our interpretation of the government's *Social Trends* publications is that the number of office workers has increased from 8.5 million to 10 million over the past 15 years, while the total working population has dropped from about 26 million to about 25.5 million' (Wharton, 1978). It is, of course, quite wrong to infer from this that 10 million jobs are likely to be directly affected by word processing.

Equally, it is tempting to combine estimates of the secretarial and typing workforce with estimates of the productivity increases attainable via word processing in order to infer the size of the automated labour pool and therefore the scope of the market. However, neither labour nor firms are homogeneous; the characteristics of typing load vary between firms and industries; and the scope for introducing word processing at all is limited by the absolute size of that load. Productivity gains made via word processing vary with the characteristics of the word processing equipment in use and those of the typing load. One user of a shared logic word processing system found, on the basis of work studies, that word processing raised productivity by factors of

2 : 1 on short documents not intended for retyping;
nearly 4 : 1 on technical documents;
and 7 : 1 on marketing letters,

as compared with a pre-existing typing pool. Overall this improved productivity by a factor of between 2 and 3. Experience at the Department of Education and Science, Darlington, showed overall gains varying between 10 and 70 per cent, according to the 'work mix', as against a typing pool partly equipped with memory typewriters (Central Computer Agency, 1978).

Clearly, word processing has a far more dramatic effect on productivity where it is introduced together with a reorganization and centralization of typing than where it is imposed on pre-existing 'best practice' in managed typing. Thus, the range of factors affecting the potential size of the market is large enough to make prediction difficult, given that so little is known about relevant intra-firm factors.

Job Content and Skills

Word processing is liable to change secretarial and typing jobs in different ways. It is generally believed that secretaries tend to spend only about 20 per cent of their time typing, devoting the rest of their time to administrative and organizational tasks. A word processor would have to be very cheap indeed to be used under these circumstances. Where offices are sufficiently large, a redivision of labour between secretaries and typists is possible, whereby secretaries specialize in personal assistant tasks and typing is fully centralized. This would have the effect of making the transition from typist to secretary very difficult and would therefore effectively remove the career path currently open to typists. The career ladder for word processor operators is likely to end abruptly at the supervisor level.

Typists, especially those working in typing pools, would

tend to have their work deskilled by the introduction of word processors. Word processors relieve the operator of many of the aspects of typing which create some variety (and introduce some 'space' into the work), such as the use of lay-out and presentation skills, as in text centring. Word processor technology is designed to maximize the amount of time which the operator spends doing that part of the typing task which the machine cannot do: operating the keyboard, principally for input of text. Manufacturers' typist conversion courses tend to be of two days duration and – whatever the manufacturers may claim – it is doubtful that the 'skill' of knowing which button to press on a word processor can compensate for the attendant loss of lay-out skills. In job satisfaction terms, this may not be a problem for operators who have learnt to operate a word processor keyboard from scratch, although absorbing some of the slightly different concepts required by electronic systems may present some problems for older people. Authors' skills are also affected. Increasingly, correspondence will be assembled from standard text components, and this could increase author productivity.

Health and safety aspects of screen-based word processing are contentious, with manufacturers, trade unions and others making conflicting claims. It is generally agreed that the environment for screen-based processing is important, that lighting should be less bright than is normal in offices, glare should be prevented, chairs should be adjustable and give adequate support, and so on (Warr and Colover, 1979; APEX, 1979; Stewart, 1979; AUEW, 1979). However, the effect on operators of flicker and electronic speeds of operation, the extent (if any) of stray radiations from the VDU cathode ray tubes and the effect of viewing emission displays on perception and concentration have been the subject of considerably more argument than experiment.

Overall, it is clear that the office is another major area where microelectronics will particularly affect women at work.

REFERENCES

Amalgamated Union of Engineering Workers, 'Health Hazards of Visual Display Units', guidance to AUEW (TASS) committee of representatives, Richmond, Surrey, AUEW, 1979.

Association of Professional, Executive, Clerical and Computer Staff, Word Processing Working Party, First Report, *Office Technology, The Trade Union Response*, London, APEX, 1979.
Charles Babbage, *On the Economy of Machinery and Manufactures*, 4th edn., London, Charles Knight, Pall Mall East, 1835.
Jane Barker and Hazel Downing, 'Word Processing and the Transformation of Patriarchal Relations', (mimeo); *Capital and Class*, forthcoming 1980.
Mary Kathleen Benet, *Secretary: An Inquiry into the Female Ghetto*, Sidgwick and Jackson, London, 1972.
Central Computer Agency, *Word Processing by Computer*, Report of a Pilot Project at the Department of Education and Science, November 1978, CCA C6.
1971 Census of Economic Activity, London, HMSO, 1975.
Department of Employment Gazette, July 1978.
The Economist, 28 April 1979.
Electronics Times, 22 March 1979.
Peter Gorb, 'Office Secrets', *The Guardian*, 27 March 1979.
Simon Nora, 'L'information de la Société: rapport à M. le Président de la République', Paris, La Documentation Française, 1978.
Ann Oakley, *Housewife*, Harmondsworth, Middlesex, Pelican, 1976.
Ted Schoeters, 'A View from the Keyboard', 'Paper May Go Out of Date', *Financial Times*, 10 July 1979.
Adam Smith, *The Wealth of Nations*, Harmondsworth, Middlesex, Pelican, 1974.
Tom Stewart, 'How to Design a Workable System', *Word Processing Now*, Business Systems and Equipment/Maclean-Hunter Ltd., London, July 1979.
Alan Warr and Joycelyn Colover, 'Visual Display Units Can Be Bad For You', *Electronic Times*, 10 May 1979.
Keith Wharton, 'Special Report: Word Processing', *The Guardian*, 6 June 1978.

PART IV SKILL REQUIREMENTS

17. Conclusions: Skill Requirements Arising from Microelectronics

P. Senker, N. Swords-Isherwood and E. Arnold

The characteristics of the manpower available to firms, in terms of the manpower resources they deploy and the availability, or shortages of skills available on the labour market, affect the realization of firms' market share, profitability and other goals. Previous chapters have analysed the competitive environment in which engineering firms have been operating, how microelectronics is affecting this environment and firms' skill requirements.

In this chapter, we attempt to draw together some of the principal general implications for the whole engineering industry.

ELECTRONICS ENGINEERS

Car manufacturers and mechanical engineering firms are short of electronics engineers, but they require both knowledge of electronics and the interest and knowledge to apply themselves to problems involving the control of mechanisms. However, most electronics engineers are fascinated by the opportunities offered by major electronics companies of working on the frontiers of electronics technology. The major electronics firms are themselves very short of electronics engineers.

Motor vehicles and mechanical engineering firms are perceived as being relatively dull and uninteresting and as offering poor career possibilities (the television industry is viewed similarly). But if some – perhaps relatively few – electronics engineers are not recruited into the mechanical engineering and vehicle industries, then these industries' prospects will be as poor as the electronics engineers anticipate. For, without the ability to develop new products and systems based on microelectronics, the problems of these industries are likely to increase. As these industries still account for over half the employment in the engineering industry, the prospects could be grim indeed. Mechanical engineering firms are using consultants increasingly to compensate for their lack of electronics engineering skills. But the use of consultants is fraught with difficulty.

MANUFACTURING ENGINEERS

This book indicates that change made in products and systems is likely to be more rapid than changes in manufacturing processes. This does not mean that change in manufacturing processes is unimportant.

The shortage of manufacturing engineers is a continuing concern. The low status of manufacturing is a problem which the United Kingdom shares with some other European countries and the United States. The need to treat manufacturing as more than a cost-cutting exercise has become more important with the increasingly rapid development of microprocessor based manufacturing systems. Manufacturing needs to be closely integrated with design.

The engineering industry is still not convinced of the need to use graduates as manufacturing engineers and it also requires some production engineers with the depth of knowledge of electronics which, generally, only graduate electronics engineers have. Such graduates, whose function may be to design manufacturing control systems embodying microprocessors, need to know about the total manufacturing process in the plants in which they work, both upstream and downstream of where they are working.

Attracting electronics engineers into such work is likely to be even more difficult than attracting them to work as design engineers in mechanical engineering. Clearly, much of the industry is going to have to rely on retraining mechanical and production engineers in electronics to meet its requirements for manufacturing engineers for some time to come. This will doubtless become increasingly satisfactory as the electronics content of mechanical engineering and production engineering degree courses increases. The need to install and run microprocessor based control systems should finally convince managements that they cannot now continue to hope for craft-trained engineers to provide their main source of manufacturing engineers.

Enhanced capability to design and install new manufacturing processes for new products will be critical, even if our forecast that manufacturing process change may be relatively slow is correct. Manufacturing engineers will need increasingly to be able to design and modify electronic test equipment to test the systems on the products they manufacture which will include sophisticated electronics. Frequently, the test equipment will be of an order of complexity greater than the product it is built to test.

TEST, INSPECTION AND INSTALLATION

The demand for test and inspection technicians is likely to grow as more and more industries incorporate sophisticated electronics in their equipment. The demand for people to do routine testing work may decrease, as modern test equipment, incorporating micrprocessors, makes possible simplified procedures.

MAINTENANCE

Numerical control machines, and more advanced systems, for example in welding, place increased demands on maintenance. There is the familiar problem of recruiting and training people with multiple skills, e.g. electronics, electrical and mechanical. It is possible, as one firm suggested, that the incorporation of diagnostics – made cheaper and more feasible by the availability of cheap, powerful microelectronic devices, could alleviate this problem to some extent. But, in the next few years, the proliferation of varieties of relatively complex machine control systems may well exacerbate maintenance problems faster than the incorporation of diagnostics alleviates them. Installation and maintenance are likely to be changed extensively by the advent of microelectronics. Where microelectronic equipment replaces electromechanical equipment, there is also substantial experience (e.g. the American experience with electronic exchanges) to show that the number of people required in such functions is reduced sharply. However, it appears that there is a relatively small though important demand for highly skilled installation and maintenance people, for example with some knowledge of software.

Where microelectronic control systems are introduced to control mechanical equipment, a few people with many skills – perhaps electronic as well as electrical, mechanical and hydraulic – are needed. Manufacturers are conscious of the need to incorporate automatic diagnostic equipment in order to simplify maintenance. The need for maintenance people will depend very much on the design resources available to manufacturers for designing automatic diagnostic equipment and on their success in designing reliable products and simplifying maintenance procedures. If they are very successful, then the need for maintenance people highly skilled across several areas will be small. But, in mechanical engineering products and vehicles, there will be an interface between the electronics and mechanical and other systems. It is, perhaps, unlikely that, however hard they try, manufacturers will be able to predict, and therefore devise automatic diagnostic equipment to monitor all the fault conditions that could occur in complex machines. Therefore, it seems reasonable on present evidence to predict that there will be an increasing need for more skilled maintenance people with multiple skills. Automatic diagnostic equipment may reduce the skill requirements, and even the numbers required, of less skilled maintenance people. This possibility of skill polarization has many possibly severe consequences. For example, it is likely to exacerbate problems of providing career progression. The needs for maintenance skills are difficult to predict and considerably more research will be necessary.

MARKETING, SALES AND SERVICE

Studies of industrial innovation (such as SAPPHO, 1971; Langrish *et al.*, 1972) have shown that firms and innovations which satisfy a perceived need in the market succeed better than those which are essentially technology driven. The function of sales is to demonstrate to potential customers that there is a connection between the firm's products and the satisfaction of customers' needs.

The industry continues to underestimate the significance of marketing and sales functions. In many sectors of the engineering industry, it is important to have marketing staff (not necessarily always the sales force) capable of feeding back and interpreting market and technical information to designers. Too often, the market research department is a haven for ageing salesmen, if it exists at all. The need for qualifications in selling depends very much on what the firm is selling and the nature of its market. The sales force needs to understand how the products they are selling can contribute to the customers' operations. To sell a complex piece of equipment to manufacturers to incorporate in their products or systems which in turn they will sell to final users for use in a manufacturer's production process (OEM sales) requires salesmen to have deep technical knowledge of the products they are selling, as well as knowledge of the customers' applications. But those selling office equipment need to understand office procedures to a greater extent than they need knowledge of the technical details of the equipment they are selling. In all cases, knowledge of selling techniques is important. In general, the increasing sophistication of the products and systems being marketed, many owing their complexity to the incorporation of microprocessors, is making marketing and sales functions more critical to firm success, and tending to increase the technical knowledge required by sales forces.

Marketing, sales and service tend to be areas of particular management weakness in the United Kingdom. Sales, in particular, suffers from a severe lack of status and a feeling that it is distasteful, even ungentlemanly, to sell hard. This particular area of British weakness tends to be one of maximum strength among Japanese and American companies, who have also realized the importance of technically expert salesmen when selling technically sophisticated equipment.

INTEGRATION OF MANUFACTURING, DESIGN AND MARKETING

In Japan, the status of manufacturing is much higher than in European or American companies. In Europe and the United States, design engineering is regarded as mainly a question of designing new products, and much technical effort is devoted to it. Manufacturing engineering is regarded as essentially a process of just 'shaving a few cents more' off the cost of a sub-assembly or chassis (Sciberras, 1979).

But in Japan considerable efforts are made to ensure that management in all functions are aware of manufacturing requirements and of the manufacturing potential and limitations of the firm. To achieve this, all trainee managers are required to spend six months of their initial two years training in manufacturing departments. All managers are trained in quality control and manufacturing management is intimately involved in the development of new products from the beginning.

Both graduates and technicians in Japanese television companies often have two years initial training, often in company technical high schools and university institutes of engineering. Following training, about a third of engineers are rotated every three or four years. This contributes to closer integration of design, manufacture and marketing than is achieved by their European and American competitors and is facilitated by the Japanese system of life time employment. Engineers and technicians in Europe expect salary increases on completion of training and on transfer to a new job, while in Japan the expectation is that the individual will do the job which the company requires him to.

ATTITUDES TO GRADUATES

Graduates are less valued in British industry than in other countries. British managers still lack formal qualifications compared to managers internationally. Experience of employing graduates is not always successful. These bad experiences are laid at the door of the graduate. It is probably true that the British graduate has unhelpful attitudes and preconceived notions about industry and those that work in it. But in Japan, in Europe and in America recruitment of graduates is considered of vital importance. Views as to the immediate usefulness of the graduate, and the amount and type of training he is considered in need of, differ, but all agree that graduates are necessary in the engineering industry.

Not all sectors of the British engineering industry reject graduate recruits. Many have active policies of graduate recruitment and could not survive without this manpower. Nevertheless, when compared internationally their recruitment tends to be less and their attitudes less enthusiastic. More dangerously we have encountered repeatedly the situation in which graduate recruitment and training has been the first area for short-term economy measures.

MANPOWER PLANNING

Manpower planning in even its simplest form is less practised by indigenous British companies than in foreign multinationals based in Britain. Large companies which do not collect basic data, for example, may find themselves with an ageing group of essential workers with insufficient time to recruit and train. One British computer company that we visited kept what little manpower information they found necessary by manual means and not on computer.

The significance of this neglect of manpower planning is both philosophical and practical. It illustrates that manpower issues are regarded as peripheral to the main objectives of the enterprise. This contrasts most markedly with the approach of Japanese companies in particular, which integrate all aspects of the planning exercise. In large Japanese companies, manpower is planned rigorously within the constraints of guarantees of lifetime employment (albeit only for men). Training is considered of vital and central importance and is not hived off to a department which has little status or power within the organizational structure. Clearly, recognition of manpower needs, particularly in the face of new technological challenges, requires analysis of the existing situation and a realization of the importance of manpower in overall planning.

The principal message of this book is that the prime requirement for competitive success and for coping with microelectronics is to have a highly educated, trained and skilled workforce at all levels. Those countries and firms which recognize this and implement appropriate policies will be successful in the microelectronics age.

REFERENCES

J. Langrish *et al.*, *Wealth from Knowledge*, London, Macmillan, 1972.

'Project SAPPHO: A Study of Success and Failure in Innovation', Science Policy Research Unit, University of Sussex, two volumes, mimeo, 1971.

E. Sciberras, 'Transfer of Technology in the Consumer Electronics Industry – The Television Sector', for the OECD, Directorate for Science, Technology and Industry Ad Hoc Group on the Transfer of Technology to Developing Countries, September 1979.

Author Index

Subject Index